SACRED DEMISE:

Walking The Spiritual Path of Industrial Civilization's Collapse,

Carolyn Baker, Ph.D.

iUniverse, Inc.
New York   Bloomington

Sacred Demise
Walking The Spiritual Path of
Industrial Civilization's Collapse

iUniverse books may be ordered through booksellers or by contacting:

iUniverse
1663 Liberty Drive
Bloomington, IN 47403
www.iuniverse.com
1-800-Authors (1-800-288-4677)

ISBN: 978-1-4401-1972-9 (pbk)
ISBN: 978-1-4401-1973-6 (ebk)

Printed in the United States of America

iUniverse rev. date: 02/17/09

ALSO BY CAROLYN BAKER

U.S. History Uncensored: What Your High School
Textbook Didn't Tell You

Coming Out of Fundamentalist Christianity:
An Autobiography Affirming Sensuality, Social
Justice, and The Sacred

The Journey of Forgiveness: Fulfilling The
Healing Process

Reclaiming The Dark Feminine: The Price of
Desire

DEDICATION

*With gratitude, honor, and reverence for all members
of the earth community, human and more than human;
to the elements of fire, water, earth, and air and to
the sacred directions: north, south, east, west, above,
below, and to the circle of life that knows no beginning
or end*

TABLE OF CONTENTS

If we do not soon remember ourselves to our sensuous surroundings, if we do not reclaim our solidarity with the other sensibilities that inhabit and constitute those surroundings, then the cost of our human communality may be our common extinction.

David Abrams, *The Spell Of The Sensuous: Perception and Language In A More-Than Human World*

FOREWORD

By Sarah Anne Edwards, PhD, LCSW

If you're reading this book you already know we're living in perilous times. By now most people are waking up to that fact. Even President Obama has warned us that difficult times lie ahead. Just days before taking office when asked if sacrifice will be required of everyone, Obama told the nation, "Everyone is going to have to give. Everybody's going to have to have some skin in the game."

But even recognizing that times are tough, chances are reading this book will feel overwhelming and unsettling at times. I know it has been for me and I've been studying, writing, and teaching about the implications of such threats as peak oil, climate change, and economic collapse over half a decade. But I'd like to suggest that we welcome whatever feelings of overwhelm or disquiet this book may stir in us, because like the medicine our mothers gave us as children, they will make us better.

I say this because Carolyn is not one to beat around the bush. Her focus is not on statistics, charts, and data demonstrating the factual realities of our eco-nomic situation. There are ample books with such information. Most of them touch lightly on the focus of this book, firmly acknowledging its importance, but skittering on to the facts.

In *Peak Everything,* for example, Richard Heinberg emphasizes that "Much of the human impact (of peak oil and climate change) will be measurable in economic terms; however, individual and collective psychological effects will perhaps be of equal and often greater significance." Having observed the effects economic collapse of the USSR in the 1990's, Dmitry Orlov agrees. In *Reinventing Collapse,* he writes, "Economic collapse is about the worst time for someone to suffer a nervous breakdown, yet this is what often happens."

Fact is, the future is going to be hard to swallow, not only presenting practical day-to-day challenges but deeply impacting our emotional and spiritual lives. In *Sacred Demise* Carolyn doesn't just touch on this fact. She dares to make it the sole focus of our attention, reaching far beneath the statistic, charts, and generalizations about their implications and delving deeply into the heart and soul of the inner traumas and turmoil we will most likely encounter. She zeroes right in on the very places we want to run away from, ignore, and rationalize, including the profound loss we will feel.

This naked honesty about such matters is only the context of the book, not its purpose. This is not a doom and gloom book. It's about looking squarely at the reality of our circumstances, as incomprehensible and uncomfortable as they may be, and exploring the potential for personal growth, transformation, and resurrection.

By openly sharing her personal experiences and the wisdom, poetry, questions, and activities included in the book, she draws upon herself to understand and grow from our circumstances by infusing them with compassion. Carolyn makes us feel safe venturing into the tender, vulnerable places.

She says let's look straight on at what we're facing and allow it to be a grand teaching to deepen our understanding of life and our place in it. While the reality of what we are dealing with may be overwhelming when seen with such stark honesty, isn't that what we yearn for? Isn't the failure to talk together in this way what unsettles us most? Like the young child whose parents won't talk about a life-threatening illness that's befallen their household or the child who on the way to the dentist is told "It won't hurt at bit," we don't want to be blindsided by realities that could be opportunities instead to prepare ourselves and make ourselves strong and capable.

After all, what is being overwhelmed but a just fast flowing river of experience we need to catch our breath for; a rug pulled out from beneath our feet to get up from, an unexpected encounter we can surprise ourselves with by responding to aptly? Chances are as you read on page-by-page you will begin to feel oddly relieved. You'll be getting your bearings in a reality that at first seems foreign and frightening but gradually reveals itself as a truth you have long known already.

Certainly that has been true for me. From the time I was a small child I sensed there was something very wrong with how things work in our human-made world. I spent many hours out of doors where everything seemed to make sense, but as I matured I concluded there must be some flaw within myself that was preventing me from understanding what everyone else seemed to grasped so easily, that our culture works to our best interest even if it appears otherwise. With that mind I undertook the arduous task of molding myself into what was expected of me hoping at some point everything would fall into place as promised.

It was not until I entered the ecopsychology PhD program that I came to see how terribly disconnected, distorted, and dysfunctional our human world has become; how tragically separated we've become from the natural world we are innately part of. As I began to learn from nature how naturally other life forms function, I was heartened by my four years in which I had the pleasure of experiencing how naturally and organically life could unfold. Oh, surely not without challenge or discomfort, but like riding down that fast-flowing river. At times it's surging furiously; at other times it's bubbling along gently. But it's always ever- changing and engaging, always keeping me on my toes, yet something perfectly natural I could navigate as long as I paid attention.

"Ah, ha!" I concluded, at last I will know how to live in our world! When I complete the program I will return to the "normal world" and just continue living in accord with nature's ways. But that was not to be. Returning to the "normal world" was one of the most difficult experiences of my life. My short-lived, free-flowing journey through life ran smack dab into a massive cultural dam of expectations and limitations as difficult for us to escape from as it is for any river to circumvent the manmade dams we've constructed to divert and contain.

This is when I realized fully how virtually impossible it is to function naturally and healthfully within the demands of our growth-oriented, consumer-driven, materialistic, and hierarchical culture. But that's the world that's crumbling now. While we may enjoy many of its conveniences and comforts, or at least have the promise of enjoying them, we also know the price we pay to pursue and maintain them is high. The stress, the long hours, the fast pace, the pressure, the fatigue, and the demands on our time by tasks far different from those we'd prefer. Our way of life is not only wearing down the natural world, it's also wearing down our psychic and our physical well-being. No matter how many possessions we garner or how much money we make, there is never enough to quell our yearning for a greater sense of peace, potency, normalcy, and well-being.

Granted, the difficulties we're facing will not bring us this kind of well-being anytime soon. Probably anything but, at first, because few of us are prepared mentally or spiritually for the onslaught of feelings, the tide of emotions, the rush of the unexpected that is engulfing us. After all, the world that anchors our daily lives now is collapsing right along with the dam upon which our current way of life in built. We don't know if we'll make it to the bottom of the spillway as the dam cracks and breaks away. Assuredly some of us won't. Even if we do, there is no guaranteed, made-to-order world awaiting us on the river once it has burst forth from the dam. For sure there will be no welcoming "comfort inns" along the shore. We're uncertain what the unleashed river of our life will look like or if we will have the knowledge, skills, and acumen to survive in it.

But life, death, mystery, uncertainty, paradox, and danger are all part of the natural world we inhabit. As Henry David Thoreau discovered and expressed so aptly after abandoning his attempt to climb Maine's Mt. Ktaadn, there is in nature "a force not bound to be kind to man, an awesome, primal force of evolving matter-in-motion," a force, I might add, that has proven to be beyond our control to manipulate and mould to our desire. This simple insight is not easily accepted in our culture where we learn that anything we believe, we can achieve, and it is something Carolyn invites us to address before it's too late.

Sacred Demise is an opportunity to do the emotional and spiritual preparation we need to be present, awake, and responsive in awesome circumstances we can't control or prevent. Each chapter invites us to delve deeper into ourselves, deeper into who we are and how we fit into the natural world outside our roles in a crumbling system. From that blatantly honest, deep place of our hearts and souls we can begin to *find* our way to a more natural way of life and glimpse how the grand unleashing of the waters constrained behind our current way of life could bring us the peace, potency, normalcy, and well-being we yearn for.

The operative word in this possibility is *find*. Throughout *Sacred Demise*, Carolyn refers to many insights from Viktor Frankl, the Austrian neurologist, psychiatrist and Holocaust survivor who wrote in *Man's Search for Meaning* of his experiences in the Nazi death camps. Some years ago I had the honor and pleasure of interviewing Frankl. His rather stern rebuke of a question I raised left me with an enigma I spent many years wondering about. Not until midway through my ecopsychology program and many learning experiences in nature did I finally understand what Frankl was getting at and why he correctly me so sharply. I had made an innocent error common to our culture, an error we all need to correct to be prepared mentally and emotionally for the years ahead.

As I began to ask how we create meaning in our lives, Frankl interrupted me abruptly, "Not *create* meaning, *find* meaning." And so it is in the adventure Carolyn invites us to undertake in *Sacred Demise*. The pages that follow are an invitation to *find* meaning in a time of collapse so that when our preconceived and manufactured dreams for how we'd like things will often no longer be relevant, it needn't be also the end of joy and value. This is a worthy task because only from a place of such personal meaning can we build new lives along life's free-flowing river, if not for ourselves then at least for those who come after us.

Sarah Anne Edwards, LCSW, PhD Ecopsychologist, provides continuing education courses for helping professionals through the Pine Mountain Institute and directs Let's Live Local, a non-profit organization working to build local resilience. She's the co-author of *Middle-Class Lifeboat,* and a trainer for the US Transition Initiatives Network.

INTRODUCTION

We white Americans of European ancestry had our indigenous cultures stolen from us, had our lands and lifestyles, our lives destroyed, just as we have gone on to steal and destroy around the globe. It happened long ago, of course, so we barely remember it…the land remembers it…it's all there, still reverberating through our bodies, like gongs still sounding centuries after they were struck with the mallet…and that pain shapes us, and hurts us still, and that, of course, is the deep reason we do what we do now to the planet. If we don't seek to understand that, how can we ever expect to change our behavior?

> Tim Bennett, writer and director of "What A Way To Go: Life At The End Of Empire"

As I stand along the surface of the Earth, she says child to me, she says grandchild to me.

> ~Navajo Prayer~

Civilization does not occur among healthy people.

> Ken Carey, *Return of the Bird People*

For reasons which I've long since forgotten, one Sunday afternoon in the year 2000 I was sitting at my computer Googling "the Central Intelligence Agency and drugs". Little did I realize that with a few key strokes I had begun a journey that would bring me to where I am today. Less than one month before the infamous 2000 U.S. election, I was beginning to scratch the surface of myriad layers of corruption in the United States government and the trajectory of world domination on which it had been traveling since at least the end of the Civil War. Everything began connecting with everything else as infinite portals of information on this topic continued to open to me like some sort of twenty-first century pandora's box. I could have ended the quest at any moment, but I chose not to. Like someone in "The Matrix" who had taken the red pill, I had to discover how deep the rabbit hole actually went. I was hooked—irresistibly drawn to delve more deeply into the information I was exploring, thirsting for even more, and within a short time would discover a quote from a former U.S. government whistleblower, L. Fletcher Prouty: "People are fundamentally suckers for the truth."

Today, with hindsight, following nine years of research and with no intention of concluding my quest, I have come to believe that while we are all *fundamentally* suckers for the truth, most of us do not want to hear it when it is the slightest bit discomforting, and tragically, most individuals living in the United States absolutely refuse to hear, see, or believe it. My research deepened after the fraudulent 2000 election, and by the time the attacks of September 11, 2001 occurred, I was extraordinarily suspicious of the official story of them provided by the U.S. government and its obsequious handmaiden, mainstream media. Although one-third of Americans now believe that the 9/11 attacks were an inside job, very few of them are willing to look deeper. I have continued to research 9/11, but a number of momentous subsequent events have overwhelmingly turned my attention elsewhere including the invasion and occupation of Iraq, Peak Oil, a fraudulent 2004 election, global warming, and at this writing, what appears to be full-blown global economic meltdown.

We are now closer to the end of the first decade of the twentieth century than its beginning, and throughout the past nine years, I have come to understand that what I and my fellow Americans and fellow earthlings are confronting is not just corrupt politics, fraudulent elections, and endless war for which in the past there may have been a political remedy, but rather, the collapse of Western civilization. As a result of my research on Peak Oil, global warming, population overshoot, species extinction, and global pandemics, I already knew that the empire the United States had become was collapsing, but when I requested and watched a review copy of "What A Way To Go: Life At The End Of Empire" from writer Tim Bennett and producer, Sally Erickson, I entered not only a more intense process of dot connection, but unprecedented reflection on what collapse would personally mean for me and every life form on planet earth.

As I pondered "meaning" not only as synonymous with "outcome" but as more intimately connected with that dimension of life which I choose to call the sacred, I was taken on yet another journey. This would not be a primarily a journey of analyzing the extent of polar ice cap melting, how much oil or water are left on the planet, how weather is affected by global warming, or how quickly I can move off the power grid and become totally self-sufficient. Instead, I would be grappling day to day, in fact, moment to moment, with the pivotal question which comprises the last ten minutes of "What A Way To Go", which is, "Who do I want to be in the face of collapse?" As a result, this book is not written as a manual of "how to survive collapse" or "how to prevent collapse" but rather, how to *be present* with collapse—how to perceive it as a phenomenon in which I am being both required and invited to participate for a specific purpose, or perhaps many specific purposes.

Collapse is not an event for which we are headed in some near or distant future. It has already begun, and we are well into it. As I state repeatedly in this book, I do not know how collapse will continue to play out, nor do I know its final consequences for life on earth or for the planet itself. What I am certain of, however, is that it will entail more loss than anyone reading these words—or the person writing them, can begin to imagine. This unfathomable loss will manifest as the loss of life, livelihoods, health, homes, marriages, families, food, clothing, shelter, towns, neighborhoods, communities, transportation, mobility, infrastructure, sanity, loved ones, possessions, and our very identities as the human beings we now are. It is likely to be the most devastating holocaust in recorded history, and it will be the end of the world as any of us has known it. *But for me, that is not the end of the story; this book is an attempt to suggest what is beyond collapse and what every person reading it may wish to consider being and doing, as collapse unfolds, in order to plant and nurture the seeds of a new paradigm for the earth community.*

Perhaps unbeknownst to the culture of empire, it has given us ample imagery for collapse, particularly with the release of "Titanic" in 1997—a blockbuster that is still riveting the imagination of the masses a dozen years later. It's all there: the hubris, the "unsinkable" titan, the division of classes, the primacy of the wealthy, the domination of women, and the soulful woman who, departing from her ruling elite milieu at the speed of light, has the courage to remove her corseted armor and pose nude for the itinerant artist who stole her heart. Fallen from her aristocratic status and drifting alone in an icy sea of death, she is ultimately rescued and lives to become the wise old woman who mesmerizes her audience with the story of the great ship's demise and seals her wizening by tossing into the sea the blue diamond which symbolized her enslavement by wealth, power, and the male tyrant that sought to possess her.

Even now the majority of those who claim to desire planetary transformation refuse to surrender to the reality of collapse and are busily re-arranging chairs on the decks of civilization because they do not understand that the current system *must* collapse. They do not comprehend that we cannot start from "right here" and reform the system, giving emergency make-overs to money and politics, thereby facilitating a glorious transformation of consciousness. In other words, they long for rebirth without the death that makes it possible. Yet even as they resist collapse, scientists like Jared Diamond (*Collapse: How Societies Choose Or Fail To Survive*), novelists like Cormac McCarthy (*The Road*), and film makers like Bennett and Erickson are captivating the minds of still other individuals who have the courage to see and talk about what is so and ponder the desirability of collapse as a necessary evolutionary trauma in the odyssey of planet earth.

In *The Long Descent*, John Michael Greer discusses the difference between a problem and a predicament. A problem, says Greer, calls for a solution, and the only question is whether a solution can be found and made to work. A predicament, however, has no solution, but faced with it, people come up with responses which may fail or succeed, but none of them "solves" the problem. Currently, humankind is embroiled in enormous predicaments as a result of problems that could have been solved or prevented decades ago. We are now staring into the abyss of the end of hydrocarbon energy on earth—a reality that was well known by presidents and petroleum geologists since the 1950s. At that time, Peak Oil was a long-term problem for which a variety of solutions could have been considered. However, as a result of ignoring the problem, it has become a predicament for which there is no longer a solution.

Moreover, as Greer notes, "Since the dawn of industrial civilization, the predicaments that define what used to be called 'the human condition' have been reframed as a set of problems to be solved.... The difficulty with all this is that predicaments don't stop being predicaments just because we decide to treat them as problems. There are still plenty of challenges we can't solve and be done with; we have to respond to them and live with them."[1]

I will not, because I cannot, write about "how to stop collapse from happening" since it is already in progress and has been well underway for decades. The title of a 2007 article of mine summarizes my perception: "The Switch Has Been Flipped: It's Too Late For Solutions." For years I have been distinguishing between solutions and options, between hope and mindset. Collapse is a fact of our lives for as long as we reside on this planet. Our work is not to prevent it but to open to it, prepare for it, and do our best to survive and live it with conscious intention and presence in relation to ourselves and all other life forms on earth as we experience it.

While I resist using it, another word for collapse might be *apocalypse*. In a recent conversation with a friend, I reiterated my belief that collapse is not only inevitable but necessary to which my friend replied, "That sounds so endtimes." I knew what she meant--rapture, Book of Revelation, Jesus on a white horse attended by thousands of avenging angels hellbent on destroying the earth. I abhor the Christian notion of endtimes with its bloodthirsty white, male, punitive god and would go to any lengths to distance myself from it. Yet the conversation with my friend later set me pondering the grain of truth in her comment. What she had introduced into the conversation was the "A" word: apocalypse.

Throughout the major spiritual traditions on earth one finds what Jung called the archetype, theme, motif of apocalypse. We are all too familiar with the fundamentalist Christian notion of rapture, tribulation, and new millennium now popularized in Tim LaHaye's *Left Behind* series. Yet Buddhists, Hindus, Muslims, and myriad indigenous traditions include, for different purposes and with their own unique embellishments, concepts of apocalypse. It appears that apocalypse is a mythic, archetypal phenomenon deeply embedded in the human psyche. Without exception, apocalypse, which actually means "unveiling" or "revealing", is perceived universally as a process in which that which is hidden will be revealed, resulting in some sort of purification. A Hopi prophecy says that "When the Blue Star Kachina makes its appearance in the heavens, the Fifth World will emerge. This will be the Day of Purification." Hopi elders believe that we are now transitioning from the Fourth to the Fifth World and that purification of consciousness is the purpose of the current upheaval.

Just as I believe that global holocaust is underway, I also believe that it will offer many humans an extraordinary opportunity for experiencing warmth, tenderness, compassion, sacrifice, sharing, authenticity, generosity, and a host of other qualities that exhibit the very best in us. However, I have also come to believe that in order to do this, we must re-attune ourselves to the sacred. One way of thinking about this is to reflect on the above quote by Tim Bennett in which he states that in the culture of civilization, those who consider themselves "non-indigenous" are at heart indigenous individuals who have had that which makes them indigenous stripped away from them. His words state explicitly that we must come to understand this and implicitly suggest that the challenge in front of us is the restoration of the exiled indigenous self.

In recent years we have witnessed the increasing popularity of the discipline of ecopsychology. Theodore Roszak states that the "goal of ecopsychology is to awaken the inherent sense of environmental reciprocity that lies within the ecological unconscious."[2] Some of the names Roszak offers for ecopsychology are: psycho-ecology, shamanic counseling, green therapy, earth-centered therapy, and re-earthing but emphasizes that "ecology needs psychology, and psychology needs ecology." A few of the issues Roszak suggests are superbly addressed by ecopsychology are: consumption habits, gender stereotyping, child psychology, environmental design, money, and the psychic need for wildness.

While Roszak prefers to speak of the "ecological unconscious", he is nevertheless, pointing to concepts that are inherent in the indigenous self which we might summarize as a fundamental enchantment with the earth and our fullest aliveness as we inhabit it. Indigenous peoples "respect" the earth not because they are "supposed to" or because their ceremony or elders dictate such reverence but because they are in love with it. *Eros,* a form of human love that the culture of civilization has twisted, distorted, debased, and on some levels entirely disowned, is not synonymous with human sexuality. While sexuality is an aspect of *eros,* so are sensuality, playfulness, appreciation of beauty, and delighting in the sounds, smells, colors, tastes, and textures of nature. When one is enchanted with a beloved, it is not possible to harm him. Rather, one seeks to protect, defend, absorb, and revel in his presence.

Richard Tarnas, author of *Cosmos and Psyche* in an article entitled "The Great Initiation" states that the culture of civilization has caused us to exchange our enchantment with the universe for autonomy. "The high cost," Tarnas states, "has been a gradual voiding of all intelligence, all soul, all spirit, all meaning, all purpose from the entire world—now exclusively relocated in the human self, through what from this point of view can be seen as an extraordinary act of cosmic hubris. This disenchantment has been discerned and lamented almost from the very start of the modern project— but what's not so readily acknowledged is that it is probably a further act of human hubris to think we were and are responsible for the disenchantment all by ourselves. There may be other, larger, forces at work." Indeed there *are* larger forces at work by way of the stories civilization has instilled in us with their attendant values of linear, reductionistic, hegemonic thinking.

In order to fall in love with the earth once again and be enchanted by it, the paradigm which civilization has inculcated must be transformed through reclaiming the indigenous self. Like so many of William Stafford's poems, "What To Do When You Get Lost" illumines this process:

> *Out in the mountains nobody gives you anything*
> *And you learn the rules after the game is over.*
> *By then, it is already night*
> *And it doesn't make any difference what anyone else is*
>   *thinking or doing.*
> *You have to turn into an Indian.*
> *You remember stories and you know that the tellers of*
>   *the stories*
> *Were part of all they said and everyone is,*
> *Even you and those all around you.*
> *But if you are afraid, you will never find them.*
> *Those questions that people used to ask: "What would*
>   *you do if…?"*
> *Have their own answer: "Nothing".*
> *Some things cannot be redeemed in a hurry,*
> *No matter about intentions.*
> *What could be done, had to have happened a long time*
>   *ago.*
> *If evil could be cancelled, it would not be very evil.*
> *The stars that you see while you drift away into the*
>   *night*
> *Have their own courses*
> *And they watch you, and they already know your*
>   *name.*

Stafford writes authoritatively about being lost and how little we know so late, and how little what anyone else thinks matters, but assures us that in order to survive, we "have to turn into an Indian." He then immediately refers to stories— not the stories of empire but those that reveal our connectedness with all that is, and implies that when we return to that connectedness—that is, our enchantment with the earth and our fellow earthlings, and when we do so with courage, we find our "tribe". In so doing, we are less lost. And then just as we sigh with a bit of relief at this point, Stafford interjects the idea of evil and how it can't be cancelled, reminding us that some things can't be redeemed in a hurry.

What could be more evil than being robbed of the indigenous self—the ultimate crime against humanity and the earth? Nevertheless, the poem ends with the reassurance that even in our lostness, there are stars that have their own courses—that is, that are not lost, who watch us and already know our names.

It is by "turning into an Indian" or reconnecting with the indigenous self that we can truly hear the stories we need to hear, find the people we need to be with, navigate the evil that pervades our world, and find our way by experiencing our connection with those aspects of the universe that are not lost—the stars, the trees, the rivers, the migratory birds and whales that always find their way as they leave their homes annually and return to them with stunning precision.

Unlike Theodore Roszak, Thomas Moore, author of *The Re-Enchantment Of Everyday Life,* who is himself a psychotherapist, defines ecology not in terms of psychology, but in terms of the sacred. "Enchanted ecology", Moore says, "is the work of religion more than science, love more than understanding, and ritual more than heroic action."[3] Ecology is for Moore "the mystery of home", and he emphasizes that if we had felt in our hearts that we needed to take care of the planet because it is our home, that emotional attachment would have prevented humans from consigning it to its present state of demise. Yet Moore does not quite capture the indigenous attitude in this statement because that attitude motivates one to take care of the earth not merely because it is one's home, but because it is literally a part of oneself—a part of one's body, one's soul, one's sacred space.

However, in the final pages of *The Re-Enchantment Of Everyday Life* Moore's words superbly resonate with the indigenous self when he states that "… once we allow the world itself to have a soul and an interior life, then enchantment begins to stir. Sensing the world's inner life, we are affected emotionally and may find a basis for a deeper connection."[4] From the indigenous perspective, the trees are not "trees" but "the standing people", that is, life forms who are, in the words of the Lakota Sioux, *Mitakuye Oyasin* or "all my relations." Thus, one cannot mindlessly cut down a tree because one is related to the tree as if the tree were a member of one's immediate family. One cannot soil the land, the water, or the air because to do so is to violate a family member.

Some readers may find words in the title of this book such as "sacred" and "spiritual" too abstract. West African shaman, Malidoma Somé, defines the sacred as the "awareness of the reality beyond the palpable world." Specifically, indigenous spirituality *is* enchantment, and as Moore notes, "Enchantment is nothing more than spirituality deeply rooted in the earth."[5] It is the way of the soul, that element of the psyche that holds the highest aspirations of the spiritual alongside our devotion to nature and our most primal human needs.

Anyone drawn to this book understands that we live in a disenchanted culture, and as Moore states, there is no way to re-enchant our lives in such a culture except by becoming renegades and planting the seeds of a new one. No doubt, if you are reading these words, you have already become a renegade and are engaged in the difficult and tedious work of extricating yourself from the values, the careers, the lifestyles, the aspirations, and the soul annihilation that define the culture of civilization.

A pivotal aspect in reclaiming the indigenous self is the experience of initiation. Most indigenous cultures have elaborate initiation procedures for their young people during the age of puberty; however, one does not have to be a member of an indigenous community to experience initiation. In *The Healing Wisdom Of Africa: Finding Life Purpose*

Through Nature, Ritual and Community, Malidoma Somé notes that "Westerners, like their indigenous counterparts, experience initiation in some form and in a constant manner throughout their life. A person who gets fired from the job faces a life-transforming challenge that must be considered initatiory….Initiation is simply a set of challenges presented to an individual so that he or she may grow." What Westerners lack is a community that observes the individual's growth and certifies that one has passed through an initiatory process. Thus, "only being part of a community will address the loneliness of modern people."[6] Much will be said in later chapters about initiation, particularly with respect to the principal initiation addressed in this book, that is, the collapse of civilization, and a great deal of attention will also be paid to community— what it is and isn't and what is required to create and maintain it.

Throughout the book I will use the word "purpose" on a number of occasions. Just as the final moments of "What A Way To Go" suggest that we ask ourselves "What is our purpose, what did we come here to do?" I will be echoing this question repeatedly with Malidoma's definition in mind, namely: "Something that the individual has framed and articulated prior to coming into a community." In his village in West Africa, special rituals are conducted to determine the purpose of the child being carried in a mother's womb, and they are based on the assumption that the child herself knows what her purpose is even before she is born into a physical body.

If *apocalypse* is an "unveiling" or "revealing", what is it that needs or wants to be unveiled? I believe that never in the history of the human race has it been more urgent for humans, awake to the reality of collapse, to understand their purpose. Our existence on planet earth at this juncture of recorded history is not accidental and is replete with purpose and meaning. Each of us individually and in community can discover our purpose and live it in the face of collapse. In fact, it may be that without collapse, we could not achieve clarity about our purpose or live it as fully as we can in a collapsing world.

As this book will argue, collapse is presenting an unprecedented, and for many individuals, an unwanted unveiling of who we are and offering a new paradigm for living intimately with the earth community. Each of us has the choice to open to and help construct the new paradigm or tenaciously cling to the dying dinosaur of industrial civilization, thereby abandoning the opportunity for personal and planetary transformation.

The paradigm that must replace civilization will be defined and implemented by each of us and all of us. The intention of this book is to facilitate your engagement in that process. My hope is that like a "power tool" in a pocket or purse, it will serve as a portable, useable implement that will support you not just once or twice but many times. Additionally, I cannot help but hope that the contents of this book will assist all of us in attentively and lovingly formulating a new paradigm birthed from the loins of our passionate enchantment with the earth and our primal aliveness as we discover ourselves falling in love with them one more time--or perhaps for the very first time.

Our re-enchantment with the earth is, in fact, a spiritual process, but it is the human soul, not the spirit, that makes enchantment possible. In an article entitled "Groundwork", psychologist and author, Bill Plotkin echoes Thomas Moore in reminding us that "We tend to think of the spiritual path as a journey that heads in one direction only: upward to spirit, to the non-dual, to the One. In the Western world, we often forget that our spiritual aspirations go downward toward soul as well as upward toward spirit. Soul and spirit, although closely related, are distinct, and the methods for approaching them quite different as are the goals of those methods." The exploration of soul, says Plotkin, leads "not upward toward God but downward toward the dark center of our individual selves and into the fruitful mysteries of nature" where lie our deeper impulses, emotions, and images. Plotkin believes that "global mind change" [paradigm shift] will require, as an essential component, a contemporary, Western, nature-based path to soul initiation—a modern version of what the nature-based people have always had.[7]

As suggested in a 1993 article by ecopsychologist, Ralph Metzner, entitled, "The Split Between Spirit And Nature In European Consciousness"[8], central to the abomination that civilization has become is a heroic, celestially-focused, earth-denying spirituality of sanitized transcendence which reveres the technical, competitive, conquering masculine principle and disowns the feminine principle grounded in nature and its non-linear mysteries. Both principles are inherent in all humans and are absolutely necessary for their optimum functioning, but civilization has run its course and is presently collapsing precisely because of a ghastly imbalance of these energies. The towering edifices of Western spirituality are now literally and symbolically crumbling, falling back upon the earth and demonstrating the folly of attempting to flee from her sacred ground.

Frequently, when I speak of embracing civilization's collapse, I am reminded by listeners or readers that civilization has given us many wondrous gifts such as the arts and certain aspects of technology that have not harmed us but only alleviated suffering. While this is so, what is also true is that what began as "gifts" have become "cultural capital" as Professor Charles Eisenstein explains in his beautiful book *Ascent to Humanity*. "…our progressive ownership of the world naturally and inevitably accompanies our progressive estrangement from the world, so that in the end we languish in the prison of me and mine which, no matter how great our possessions, is far narrower and dingier than the unbounded Wild from whence we came." [9]

Indeed civilization has wrought astounding wonders, but as we acknowledge that, we must also ask, at what expense? How is it that so many of the "blessings" of civilization are owned and managed by private interests for their own profit? Moreover, why does the "starving artist" or musician struggle to survive while pursuing her craft in the milieu of a society that rewards athletes with obscene salaries and corporate CEO's with unconscionable bonuses for essentially doing very little work except transferring money from one place to another?

As a result of civilization's disownment of the indigenous self, few individuals in Western society are willing to face the reality of collapse. Native peoples worldwide have spoken and written about collapse for millennia, but heroic modernity refuses to recognize its irrevocable reality. From the heroic perspective, either collapse "can't" happen or it is decades away or humankind will discover the magic-bullet techno-fix to prevent it or a "savior" political candidate will be elected who will nip it in the bud or "green" politics and "green" shopping will allow us to have "fun" as we save the world with even more consumerism.

Currently, those individuals who are tuned in to the reality of collapse, myself included, are perplexed, disappointed, frustrated, sad, frightened, and angry as we observe that at least 99% of the population of the United States does not perceive the current configuration of world events as the collapse of civilization or is absolutely unwilling to entertain the possibility that civilization ever *could* collapse. In her blog post "Horror Movies And Other Things I Don't Want To Believe Are True,"[10] Sally Erickson reveals that after touring the U.S. with Tim Bennett, she has realized that very few people in America want to know the truth about the state of the planet. The tone of her lament echoes my daily experience as I witness students, friends, and most Americans determined to remain oblivious to reality, blithely ensconsed in infinite layers of denial. When the issues are presented and individuals appear to listen to them, one is taken aback with the almost knee-jerk response of "I don't want to hear the problems; tell us about solutions. Tell us what to do."

I have come to understand that in most cases, this question originates not from genuine comprehension or concern for making change happen, but rather the desire to evade the gravity of the situation through meaningless, perfunctory action that makes little or no difference so as not to look more deeply into the frightening abyss of collapse. Those who take collapse seriously and are working to prepare for it are often called fearmongers, as if being fearful regarding the end of the world as we have known it and the likelihood of mass die-off and extinction and the destruction of the ecosystems is somehow an inappropriate response. The accusation implies that the entire truth about these realities should not be told, and that if they are, the bearer of bad tidings should be ashamed of him or herself.

Whenever I encounter this response I remind the accuser that some years ago Gavin DeBecker wrote *The Gift Of Fear* to instruct humans regarding the survival value of that emotion—a natural, human response to threat of harm. The author emphasizes that rather than discounting our fear or shaming ourselves for feeling it, we need to pay attention to it because its purpose is to warn us of impending danger *so that we may take action to protect ourselves.* One the one hand, we can allow fear to paralyze us, or we can utilize its presence in our physiology to empower and motivate us.

When attempting to educate and dialog about collapse, individuals frequently state that they don't want to know more about collapse because they don't want to spend their lives worrying about it. Such a statement is based on the assumption that if I know more about collapse, it will consume my life, causing me sleepless nights and a stress-ridden life. My experience is quite different, as is that of a number of close friends who have been learning about collapse for longer than I have.

Worrying about something is both immobilizing and debilitating; therefore, worry is not useful, much less empowering. The fundamental gift of fear is its capacity for motivating us to respond. As stated above, we cannot prevent collapse or "fix" the issues that have engendered it, but we can take action in our own lives on a number of levels.

I notice that since I have stopped denying the reality of collapse or trying to avert it, enormous energy has been freed within me to respond in ways that feel necessary and appropriate for me. I also notice that my creativity has been enhanced exponentially. I do not spend a great deal of time "worrying" about collapse, but rather taking action to prepare myself, as well as enjoying activities that bring me peace, conviviality, fun, joy, humor, and play with friends. Thus, my experience with letting the full force of collapse into my body and mind has ultimately been one of expansive empowerment. In fact, I can authentically report that I have never lived a richer, more fulfilling, more rewarding life than I'm now living—even as I learn more every day about collapse. More succinctly, collapse gives meaning to my life, and my life gives meaning to collapse.

At the same time that collapse offers meaning and purpose to one's life, the inescapable truth about it is that a thorough understanding of it requires of any aware, responsive person, broad and sweeping changes in his/her life. Specifically, the "What A Way To Go" website suggests five:

1) Fully acknowledge and internalize that the culture of Empire is destroying the support systems on which the community of life depends, and robbing us of our essential humanity.

2) Talk about your concerns with everyone you know. Make Peak Oil, climate change, mass extinction and population overshoot household words.

3) Find your work in the world to preserve life, change this culture and /or create restorative ways for individuals and communities to live in harmony with each other and the non-human world.

4) Assess what you actually need during this transition in order to live and do your work. Only buy what you need and buy from local sources in order to support the creation of local economies.

5) Find or deepen your spiritual connection to that which is greater than you. Ask and then listen for guidance about how to live joyfully and creatively in the face of these unprecedented times.

The first thing one notices in reading these suggestions is that they require one to stop—yes, I said STOP. A thorough understanding of collapse means that life cannot go on as it has in the past; business as usual is over. Furthermore, the changes suggested are enormous. If these seven sentences are taken seriously, then everything in one's life must change—dramatically. Change, for even the most sage among us, is stressful and uncomfortable, and these changes require an enormous amount of work. Who would choose to pursue this path when it is much easier to pretend that life will go on as it always has, that techno-fixes will be found, that a dream political candidate will initiate a mass movement to prevent collapse, or that changing light bulbs or installing solar panels on one's roof will make everything OK?

While encompassing all of the suggested five changes, this book will focus more specifically on Numbers 3 and 5, and it will offer definitive tools to facilitate navigating them. At the end of most of the chapters a series of reflection exercises appears which invites the reader to ponder the chapter's contents in a manner that is relevant to his own experience. The exercises have been designed to provide a map for navigating all five of the suggestions, offering a degree of structure but open enough for application to each reader's unique reality. The ultimate intention of the reflections is to assist the reader in answering her own "Who do I want to be?" question.

Alternating with left-brain prose in this book is a variety of poems which I have cherished for many years because of the numerous occasions on which they have deeply touched and guided me when nothing else could. I invite you to collect your own treasure-trove of poems, songs, stories, and objects of art and allow them to speak to you alongside all the horrors regarding the demise of the earth with which we are daily bombarded. Unlike those who would tell you to "stay positive" or those who have long since become overwhelmed with nihilistic despair, I would encourage you to feel all of the natural, sane, responses to the end of the world as we have known it, including grief, terror, rage, shame, and yes, despair, and at the same time, nurture yourself with all that feeds your soul. Otherwise, you are living only one half of the whole story of collapse. The story you are living in this moment is unique in human history, and all good stories are comprised of the very worst, the very best, and everything in between.

Peter Senge, founder of the Society for Organizational Learning and author of *The Leader's New Work* honors the power of creative tension resulting from "seeing clearly where we want to be, our 'vision,' and telling the truth about where we are, our 'current reality'." From holding vision alongside current reality, creative tension emerges which allows for the dynamic realization of possibilities which could not have been created by fixating only on current reality or only on one's vision. "Without vision," Senge says, "there is no creative tension. Creative tension can't be generated from current reality alone." Often we remain in analysis of the current situation to our detriment because as Senge notes, "All the analysis in the world will never generate a vision."

Conversely, says Senge, "...creative tension can't be generated from vision alone; it demands an accurate picture of current reality as well." This is one reason I cringe when people talk about "preventing collapse" or refer to those of us who are willing to look deeply into the abyss as "doomers." They appear to insist on having the groovy, green, good times rolling endlessly or bypassing feeling all of the painful emotions that collapse quite naturally evokes by quickly supplanting them with their vision. Conversely, it is equally true that staring into the abyss without vision may be equally unproductive.

In fact, Senge reminds us that:

> The principle of creative tension has long been recognized by leaders. Martin Luther King, Jr., said, "Just as Socrates felt that it was necessary to create a tension in the mind, so that individuals could rise from the bondage of myths and half truths, so must we create the kind of tension in society that will help men rise from the dark depths of prejudice and racism."

But King was assassinated, and only part of his vision has been fulfilled. In the case of the collapse of civilization, cataclysm could result in the extinction of humans, and subsequently, millennia may be necessary for the earth to restore itself. So why have a vision? After all, my vision may be nothing more than a pipedream; if so, what's the difference between having a vision and taking Prozac?

I would argue that even in the face of extinction, holding the vision is necessary for a holistic perspective that embraces more than current reality alone. Moreover, creative tension is only available *as a result* of holding the vision.

I further argue that none of us is an "accidental tourist" in the saga of collapse. While it offers us absolutely no choice to avoid it in this lifetime, it does offer us an option regarding *how* we will meet it. There are only three choices: We can either opt for death, denial, or deliberation.

Opting for death does not mean suicide or developing a psychosomatic terminal illness; one can manage to survive collapse and still choose death by enduring it as nothing more than meaningless suffering. Yet it seems that, the majority of earthlings appear to be opting for death by way of denial. However, opting for deliberation is, in my opinion, synonymous with choosing life, which does not mean clinging to the soporific of "hope"—

yet another face of denial, but rather, exploring the inner recesses of the soul and pondering what collapse means and what ultimately matters. This book has been written specifically for the purpose of providing a structure for choosing deliberation and introspection—not narcissistic navel-gazing, but deep, conscious contemplation of collapse and its emotional and spiritual implications for you, the reader.

Walking the spiritual path of collapse is a journey that beckons us far beyond mere survival and renders absurd any attempts to "fix" or prevent the end of the world as we have known it. This odyssey is about the transformation of human consciousness and the emergence of a new paradigm as a result discovering our *purpose* in the collapse process and thereby coming home to our ultimate *place* in the universe. Our willingness to embark on the journey with openness and uncertainty offers an opportunity for experiencing the quantum evolutionary leap with which collapse may be presenting us. In other words, the opportunity collapse offers is an extraordinary death/rebirth phenomenon which could be tragically aborted if we persist in denial, rationalization, or resistance to civilization's demise.

In an earlier version of this book, I subtitled it "restoring life on a dying planet." Restoration can only result from a paradigm which diverges dramatically from the paradigm of civilization. It is unlikely that most people alive today will live to witness the fruition of life restored on a dying planet. Perhaps the most any of us can do is plant the seeds. If we accept that that may be so, then we are immediately in the territory of the sacred because our vision will have been extended beyond mere physical survival.

Restoring life on a dying planet requires a larger vision than enhancing the longevity of oneself and one's loved ones. It compels us to consciously and compassionately re-engage with the earth community by breathing new life and love into every species of animal, plant, and mineral in our world, whether we personally prevail or not. However, that will not be possible unless we surrender to collapse and allow it to remake us so that subsequent generations of earthlings will not repeat the madness that has resulted in the annihilation of our habitat.

The journey through this book will not be easy, but collapse will be much more painful, arduous, demanding, draining, and frightening than anything you might read here. In fact, working with this book could lessen some of the severity of collapse for you personally. What is certain, however, is that collapse is happening, it appears to be out of our control, and regardless of how we logistically prepare our bodies, minds, homes, families, and finances for it, the emotional and spiritual challenges it will present to us, behoove us to embark on the contemplative process this book offers.

As I surrender this work to the printing presses, I have no words to express my deepest, warmest gratitude and blessings to Tim Bennett and Sally Erickson, Kathleen Byrne, Eckhart Tolle, Thomas Moore, Sobonfu Somé, Mary Oliver, William Stafford, Malidoma Somé, Joanna Gabriel, Suzanne Duarte, and my entire new Vermont family for their light in my life and work. Some of you I know; others, I have never met. Nevertheless, your words and work have compelled me to write this book and include you in it. And you dear reader who has intentionally or unintentionally opened this book, I support you in joining all of us in looking deeply at the realities that we'd rather ignore. I invite you to consider that the time for "ignore-ance" is over; we no longer have the luxury. What is more, you are not alone; you do have allies—some that you know about and others whom you have not yet met. Please join me and the other voices who have participated in shaping this book in creating and nurturing a new paradigm for humanity and for the earth community.

If you put your heart against the earth with me
In serving every creature, our Beloved will enter you
From our sacred realm, and we will be, we will be
So happy.

~Rumi~

THE FOUR DEADLY MYTHS

Originally Printed in *The New Hampshire Gazette*[11]

Earth has self-regulated and balanced itself for over 3,500 million years, sustaining an environment that has allowed life to flourish and thrive. This balance is now being dangerously altered and disrupted by the modern human species who has decided to supplant this natural system with a set of four deadly myths.

Scientist James Lovelock's 1979 Gaia Hypothesis, revealed that the earth, itself, is a living entity, responding, adjusting and co-evolving amid constant change, maintaining a stable, life-sustaining balance. Early humans lived connected to this life-force, in harmony and with reverence for its abundant gifts. They did not significantly alter, destroy, toxify or over-harvest, leaving enough to re-generate for future generations.

Daniel Quinn's book, *Ishmael*, tells the story of how, about 10,000 years ago, humanity made a drastic shift. Humans, called "Leavers" had lived in balance with the earth system for millions of years. At that time, others, referred to as "Takers," broke away, believing they held dominion over the earth and did not have to live within its non-negotiable rules. Today the world is dominated by "Takers."

Takers decided to live outside of nature. To do so, they created four self-destructive, suicidal myths.

Myth #1: The human species is distinct from the natural system and does not have to live by its laws. Fact is, we are intractably interwoven into the fabric of all life. All of our actions have consequences.

Myth #2: Through human "wisdom" and creativity, the forces of nature can be tamed, harnessed and controlled. History has proven this to be laughable, except for its dire consequences.

Myth #3: The economic growth and technological "advancement" of the "civilized" world creates a "better" life. This "enhanced existence" however, requires the degradation and annihilation of natural systems for the benefit of a few, self-selected humans.

Myth #4: Growth can continue unabated in perpetual "endless more;" there are no ecological limits but, even if there are, technology will "solve" any problems we cause.

These myths form a convenient veil for the corporate mantra of "endless more," which has resulted in rampant, destructive consumption of earth's resources, ubiquitous toxification and the unprecedented killing of habitat and species. The 1972 book Limits to Growth, illustrates the consequences of this illusory, self-destructive behavior. Our planet has distinct limits in its ability to sustain life, feed and shelter people and absorb waste.

Today, we are, in fact, living precipitously beyond those limits.

The corporate/governmental deception, seduces many into the trap of endless-more consumerism keeping people complacent, conveniently uninformed and shielded from the destruction and unsustainability of its ways.

Corporations, the ultimate "takers," are creating enormous, stress on the Gaian balance.

Despite our "dominance," most humans are incredibly vulnerable, entirely dependent on man-made, energy-gluttonous, industrial systems for their survival. Unless humans release their suicidal romance with these four deadly myths, and re-connect to the earth, we can expect two potential scenarios to eventually play out. One, the human-made system will collapse, jeopardizing the lives of millions who are dependent on it for food, warmth and shelter. Two, Gaia's self-defense mechanisms, capable of shifting precipitously to restore balance, will. Just as our bodies send out anti-bodies to kill invaders, the Gaian force will somehow respond to the human virus threatening her life.

It will not be pleasant.

ON CONSCIOUSNESS AND COLLAPSE PREPARATION

I see there's a big danger in making consciousness change the top thing or the priority. It's going to take a lot of physical work, social changes, economic changes, many sorts of changes which are going to require work, adjustments and decisions, in the active world. Those things are going to happen more fluidly, more easily, and I think, more constructively, if at the same time we are freeing our consciousness up from the old mode, that old, self-centered, materialistic mode, which actually created the problems in the first place on a larger level.

If we don't also tend to our consciousness, our mind, our psyches, then we're going to be repeating some of the same mistakes. I think the opposite error is equally important to avoid—which is purely looking at what must be done, without bringing in the consciousness element. That's the way we normally approach things— we see a problem out there, we must fix it, change it, do something about it. We have to bring in the mind— what is it in our thinking that's causing our problems or stopping us solving them in the right way?

So there definitely needs to be a balance, the two supporting each other, but a shift in consciousness is going to support the work we have to do on the outer. I don't believe that if we all just meditate and sit back, the world will miraculously sort itself out. I don't believe that at all. It's definitely going to need a lot of very challenging work, challenging decisions, which is going to probably push through some personal discomforts, hardships maybe, as we make adjustments. That's not going to be easy, and that's why I think we're going to need to be looking at our consciousness at the same time so that we can ease our way through that process.

Peter Russell, author of *The Consciousness Revolution* and *The Global Brain: Awakening Earth for a New Millennium*

CHAPTER 1–WHAT IS COLLAPSE?

Politics is your spirituality demonstrated.

The New Revelation

~Neale Donald Walsch~

In 2005 Jared Diamond released his provocative book *Collapse: How Societies Choose To Fail Or Survive*—a work that put the concept of the collapse of civilizations in the faces of the world's inhabitants. The most disturbing aspect of Diamond's book, aside from the fact that no one really wants to talk or think about collapse, is that he asserts that societies are not murdered but essentially "commit suicide." But before we can fully comprehend that notion, we must ask: What exactly *is* collapse? Does it mean total extinction? Does it happen slowly or quickly? What are the signals that collapse is about to happen?

My dictionary says that collapse means "to fall or cave in suddenly", "to break down, to come to nothing", "complete failure", "disintegrate", "crumble", "to sink into extreme weakness." While the first definition adds the word "suddenly", there is nothing about the others that implies rapidity. In fact, most of them imply a process that moves more slowly. More recently, author John Michael Greer, in *The Long Descent: A User's Guide to The End Of*

The Industrial Age, asserts that, "Like the vanished civilizations of the past, ours will likely face a gradual decline, punctuated by sudden crises and periods of partial recovery. The fall of a civilization is like tumbling down a slope, not like falling off a cliff. It's not a single massive catastrophe, or even a series of lesser disasters, but a gradual slide down statistical curves that will ease modern industrial civilization into history's dumpster."

Many would argue with Diamond, Greer, and with me, and insist that collapse is not happening at all and won't for centuries or millennia. They might demand footnotes and documentation for the assertion that civilization is heading implacably toward collapse, or they might accuse those who write and speak about collapse as "apocalyptics", "dystopics", "doomers", or "end-time fanatics." From me, they will not receive the elaborate documentation they might prefer because although I am an academic and accustomed to scholarly writing, these pages are not an attempt to convince the reader that collapse is real and is actually happening; I embraced that reality long ago and feel no need to defend it. Furthermore, I suspect that if you are reading these words, I don't need to convince you of the reality of civilization's collapse which is now ubiquitous.

For those who absolutely demand "proof", I would recommend the documentary "What A Way To Go: Life At The End Of Empire" written and produced by Tim Bennett and Sally Erickson which is replete with hard evidence from a variety of scholars that climate change, Peak Oil, mass extinction, population overshoot, and global pandemics are not only real but bearing down heavily upon us. Ironically, "What A Way To Go" opens with archival footage of a man standing on a window ledge several stories above a city street, contemplating jumping to his death. The suicidal man shows up several times throughout the documentary, juxtaposed with the film's fundamental message that humans at this juncture of history, which may in fact be the *end* of history, can opt to continue committing collective suicide, or we can open to the transformation that civilization's collapse is inviting us to experience.

For me, the evidence is in, the discussion is over. Collapse is now a topic with which I am intimate—I live it, breathe it, write, speak and think about it dozens of times a day. But that was not always the case. Like millions of other individuals, I didn't *want* to think about it. I preferred to immerse myself in progressive politics—obsessing about the next viable political candidate, the next antiwar protest, the next alternative energy solution, the next petition that needed to be signed. I have since come to understand that no politician, no green movement, no political party, and no amount of protest can mitigate civilization's collapse, particularly when so few individuals will even discuss the topic—a topic they are unlikely to seriously consider until they are reeling in its throes.

Three years ago I began exploring the issue of collapse. Since then the velocity of collapse has accelerated faster than I could have ever imagined. And now, in 2009, as banks fail, as avalanches of foreclosures, bankruptcies, and massive unemployment catapult millions of Americans over a financial cliff; as more Americans than ever in our history are drowning in debt and relying on food stamps to feed them—and as food banks are experiencing unprecedented shortages; as Peak Oil drives energy prices up and down cliffs and valleys of fluctuation and will eventually bring about dramatic shortages; as a evaporating pensions and 401K's erase what little security the working and middle classes had left; as climate change, the insane use of ethanol, and the privatization of water hurl the planet ever closer to global famine and thirst, I have absolutely no need to argue about the reality of collapse. I am merely watching it unfold as a host of individuals around me begin to understand that my prognostications about collapse, rather than founded in delusion, were exceedingly accurate. None of this means that I am psychic or smarter than anyone else, but that I was able, as are many readers of this book, to move beyond denial and read the glaring, indisputable indicators of civilization's demise.

I have written in *U.S. History Uncensored: What Your High School Textbook Didn't Tell You*, and I have taught for years, that corporations run the world, as David Korten argued in his 2001 book[12], and that as a result, a presidential candidate cannot even receive a nomination from his party, let alone be elected unless he is solidly in the pocket of the corporate interests of the United States. Moreover, overwhelming evidence indicates that fraudulent electronic voting technology threatens to eliminate legitimate elections in America. Given these two realities, I have lost all faith in the viability of my government and its political process. I no longer live in a democratic republic, but rather in an empire--a *corporatocracy* which is becoming increasingly totalitarian and disdainful of the earth and its inhabitants, save those of its own ruling elite cabal. Clearly, the magnitude of the collapse that is beginning to rattle our walls and shatter our windows is beyond anything governments, politicians, movements, or individuals can prevent—larger than any phenomenon in the history of the planet itself.

At this point, the reader may be wincing with "what to do?" questions or may have already put the book down. We Americans, more than any other earth dwellers, seem compelled to demand directions for taking action. "What do we do about this?" the reader says. In reply I would answer that there is almost nothing we can do *about* this, but there is much we can do *with* it, and the latter is the crux of what this book is about—not how do I prevent or avoid collapse, but how to I open to it, contemplate it, prepare for it, embrace it, and allow it to become my teacher? In his article "Money And The Crisis of Civilization", Charles Eisenstein says, "Individually and collectively, anything we do to resist or postpone the collapse will only make it worse. So stop resisting the revolution in human beingness."[13] *In other words, collapse is the next step in our evolution, transitioning us to the fullness of our humanity.*

This book suggests that the suicide which civilization is committing, like all suicides, has meaning and as the psychology profession would say, is acting out something terribly tragic—something permeated with pain, rage, despair, hubris, terror, and anguish. It is the natural tendency of any sensitive, compassionate human being to want to intervene to prevent suicide, but in this case, we have not done so, and our most heroic current efforts are not thwarting it. Perhaps something else is being asked of us.

At the age of 28 I experienced the suicide of a close friend with whom I invested a great deal of energy in demonstrating kindness, compassion, empathy, and many hours of listening. I was not the only such person in my friend's life. Nevertheless, she took her own life at the age of 27. It was one of the most profound and life-changing experiences I've ever had. To this day, 35 years later, I occasionally think about my friend's suicide, but never without recalling how it altered my life. It was the first time in my adult life that I struggled with questions of meaning such as: What is death? Who was my friend? Who is any of us? What is the soul? What happens when we die? Although I could not "save" my friend, and have never really felt guilty about not being able to do so, I will always reflect on her death as one of the most significant experiences in my life. In death, more than in life, she has become my teacher, but that could not have occurred, had I repressed the memory of the experience or avoided reflecting upon it or refused to contemplate its meaning in my life. In other words, I chose to welcome its presence and in a sense develop a relationship with the experience rather than sending it away.

I believe that the suicide of a collapsing civilization is offering us a similar choice. Will we go to any lengths to deny its reality—distract, shop, numb ourselves with addictions and activities and the heroics of ineffectual political and social movements, or will we stop pretending and open ourselves to the opportunities that collapse offers us? For me, that is the essence of all spiritual practice, namely: What do I do *with* anything rather than what do I do *about* it?

Many people ask me what I think collapse will look like. The truest answer is that I really don't know, but scientists and scholars of energy depletion, overpopulation, global warming, epidemiology, and other disciplines related to collapse phenomena are giving us disturbing clues. Partially as a result of human nature and in part because of our denial, we tend to put collapse and its consequences in the future. But I have found this to be an irrational and inefficient means of approaching the issue. What I prefer instead is to explore how collapse is looking *now* because it is not something down the road; it is here and now. In fact, I have to wonder how historians, if any are alive on earth in one hundred years, will date collapse in their historical narratives. When did it "begin"? Did September 11, 2001 mark its inception? Or was the process more subtle, more protracted?

Most of us know what collapse is beginning to look like in 2009. Every day the endangerment or imminent extinction of another species, if not several, is announced. We are incessantly reminded of melting polar ice caps, the plummeting of gargantuan chunks of glacial ice into the sea, the rising of ocean levels, dramatic climate change, ravaging fires in places where they rarely occur and more vehement fires in places where they do occur, new threats of global pandemics and previously-unknown viral strains, staggering human population statistics, droughts and the realization that along with having almost no unpolluted air to breathe two decades from now, we will have enormous difficulty securing clean water to drink—if we can do so at all.

But those manifestations of collapse are blatant and electrifying. What about the more insidious indicators? What about massive departures from organized religions and Catholic diocese bankruptcies worldwide as a result of child sexual abuse litigation? What about fundamentalist Christian clergy who move in and out of the sexual orientation closet like chameleons depending on what is most convenient and politically advantageous?

What about a colossally dumbed-down educational system that produces valedictorians who cannot write a complete English sentence or solve a simple arithmetic problem or who experience reading a book with the same pleasure they might experience in having a root canal? Is an educational system that can only produce standardized children by forcing them to take standardized tests five hours a day, four days a week, functioning in anything but a state of abject disintegration?

What about a health care system that is so broken and inequitable that it doesn't even deserve the name "health care"? Do not ever-burgeoning statistics, at this writing about 50 million Americans walking around without health insurance, and several million more who have been forced into bankruptcy as a consequence of catastrophic illness, as well as the reality that at this moment, the American employment market is being precariously propped up by jobs in healthcare—does not all of this proclaim loudly that the U.S. health care system is in shambles? As for mental health care in America, for at least the past decade it has deteriorated to the point where few individuals can pay out of pocket for mental health treatment, and if they have insurance, are limited by their providers to a designated number of covered sessions per year.

As ever-new political scandals perpetrated by corrupt politicians erupt on the scene almost weekly or monthly; as America's infrastructure rots and bridges collapse and the nation's prison population soars to an all-time high and exceeds that of every other nation on earth; as peak oil, peak natural gas, peak food, peak water, peak air, and as Richard Heinberg writes, "Peak Everything"[14] scream their collective chorus of resource scarcity, does any sane human being dare proclaim with a straight face that collapse is not real—that those of us writing, talking, and agonizing about it are hysterical gloom-and-doomers? If the disintegration and unprecedented demise of every institution in the United States of America does not portend collapse, then what does?

Yet I have not mentioned the horrific notion, perhaps likelihood, that in a collapsing world, nuclear exchange or holocaust are ghastly possibilities as everything that humans treasure seems to crumble around them. Nor have I pointed out that in the waning of what have appeared to be nearly invincible institutions, unprecedented levels of violence are likely to erupt and spread. A people as unprepared and steeped in denial of collapse as most of America, will probably become nothing less than hysterical as the crumbling exacerbates. What has someone who has lost his home, his job, his healthcare, his life savings, his pension or 401K to lose by forcefully taking the food he and his family need to eat? Unquestionably, a society in the throes of collapse, like that of the Great Depression, will manifest high numbers of suicides, explosions in the rates of addiction, and stress-induced psychoses and post-traumatic stress disorder. In a nation with more firearms than any country on earth, what might be the consequences of masses of hungry, dispossessed, homeless, unemployed, terrified human beings who have become panic-stricken and desperate? And, how will law enforcement and the military address such chaos?

Inherent in the crumbling of U.S. political and governmental institutions is the unprecedented deterioration of fundamental Constitutional liberties during the past decade—a reality of which the majority of Americans are not even aware. Simultaneously, private security companies such as Blackwater and others have been used not only to maintain order alongside U.S. troops in Iraq, but were also used after Hurricane Katrina in New Orleans in 2005. These private mercenary armies are accountable to no one and are not bound by the U.S. Constitution. In fact, in 2007 the Pentagon announced that it is conducting simulation exercises in specific American cities in preparation for possible chaotic scenarios resulting from climate change, a nuclear attack, pandemics, or natural disasters. Clearly, law enforcement and the military are anticipating widespread lawlessness and the possibility of dealing with an unruly citizenry. The possibilities of martial law, suspension of the Constitution, a full-blown dictatorship, and immediate imprisonment for dissenters or generally unruly individuals are daunting.

No wonder we don't want to think about the reality of collapse and the full extent of its consequences. And yet, as psychologist Carl Jung emphasized, whatever we deny or pretend to ignore does not go away, but only becomes larger in its power and influence within the psyche and in our external world. In not attending to the reality of the collapse that is already happening around us and that can only exacerbate, we risk not only encountering it with a lack of pragmatic preparation, but we are likely to deprive the deeper, intangible, eternal layers of ourselves, which some people have called "the soul", of the meaning that societal and ecological disintegration may offer us.

Victor Frankl, the famous physician who survived a Nazi death camp and wrote *Man's Search For Meaning*, stated that the camp inmates who every day could wake up and find some meaning in the horror of their experience were more likely to fortify themselves sufficiently to maintain the stamina necessary to simply endure. The collapse of civilization portends at the very least, massive, widespread loss. It will affect every aspect of our lives—physical, emotional, mental, and spiritual. In my opinion, a perspective that embraces meaning is pivotal in confronting the aspects of collapse that are already occurring and are likely to intensify. This book is written in part to assist the reader in cultivating intimacy with meaning while navigating the emotions and events that collapse is likely to foist upon us.

Chapter 2–Paradigm Shift— Beyond Capitalism or Political and Economic Systems

A wonderful thing happens when you give up on hope, which is that you realize you never needed it in the first place. You realize that giving up on hope didn't kill you. It didn't even make you less effective. In fact it made you more effective, because you ceased relying on someone or something else to solve your problems—you ceased hoping your problems would somehow get solved through the magical assistance of God, the Great Mother, the Sierra Club, valiant tree-sitters, brave salmon, or even the Earth itself—and you just began doing whatever it takes to solve those problems yourself.

~Derrick Jensen, "Beyond Hope", Orion Magazine, May/June, 2006

In 1962 a graduate student in theoretical physics named Thomas Kuhn published *The Structure of Scientific Revolutions* in which he introduced the idea of the paradigm in relation to the scientific community and articulated the concept of the paradigm shift. Essentially, a paradigm is a model or set of concepts and values embraced at a particular time by a particular community. However, as new ideas are entertained and divergent concepts evolve, the paradigm is altered at first gradually, but eventually as a "revolution", whether a political one, or as a revolution of ideas.

In thinking about collapse it is crucial to understand that collapse is unfolding as a result of specific paradigms that humans have held for centuries that are resulting in the destruction of the human and non-human realms. The prevailing paradigm of civilization is that humans are superior and are entitled to dominate nature and its bounty which humans identify as "resources." One of the most outspoken critics of civilization, Derrick Jensen, says that because humans have not developed a relationship with the myriad facets of the natural world, we have come to call petroleum, water, forests, arable land, wildlife, and other natural phenomena "resources." Inherent in the prevailing paradigm is a sense of entitlement that fosters an attitude of "ownership" in relation to nature and the non-human world.

The earliest humans were hunter gatherers. They did not "grow food" because food grew. They hunted and gathered it. This forced them to be nomadic and on the move all the time. Babies were carried on their mothers' backs as the tribe wandered in search of food. Although the lifespan was not long, this contact with the mother produced healthier and more secure children.

Hunter gatherers were by and large indigenous people who created a natural culture or a culture based on maintaining a friendly relationship with nature and with each other. Hunting and gathering food required humans to develop a very intimate relationship with nature and with other human beings. Therefore, the essence of natural cultures is cooperation.

Eventually, humans developed agricultural societies where they grew their own food. This had at least two effects: 1) The land was disturbed and all of the current problems we have with agriculture began—depletion of soil being one of the main agricultural issues, 2) No longer nomadic, people became sedentary. As a result, life spans and population increased. In order to feed populations and remain sedentary, they had to grow more food which produced more soil depletion and thus "grew more people". Hence, a vicious circle of soil disturbance and overpopulation resulted.

When people become sedentary, they become dependent on the land to grow their own food. When they deplete one area of land, they have to move to another area and start all over again, invariably encountering other tribes or groups that are doing the same thing, and the result is competition for land and resources, culminating in wars of conquest.

Also, when people become sedentary they establish empires, that is, political units established by a king, a royal family, a dynasty, or a small group of wealthy individuals who control the resources of the society and must constantly spread their influence by conquering more territory and resources in order to continue to exist. The most important thing to understand about an empire is that it needs endless war in order to survive *and* that the lifestyle and culture of empire is unsustainable. An empire must reject a culture of nature and establish its own kind of culture of domination and control—a culture of hierarchy in which wealthy males are at the top of the food chain and nature, women, and the less wealthy are at the bottom.

Throughout history there have been countless empires. Examples are the Greek, Roman, Ottoman, British, Chinese, Japanese empires, to name a few. Because empires are unsustainable, they always—without exception, come to their demise. Furthermore, no matter how much an empire may be attacked or overpowered from the outside, it always dissolves because of internal decay. In other words, empires always implode from within.

During the 16th century, after centuries of control of Europe by the Catholic Church, people began to reject religion as the dominating force, and European countries which had overall been ruled by religion, became secular, fostering curiosity about scientific and geographic exploration. Additionally, these countries became empires which sought conquest of land and resources.

The most famous examples of imperial countries were: Spain, England, France, the Netherlands, and Portugal.

They all explored and conquered portions of the Western Hemisphere, North, South, and Central America and established colonies there in search of gold and other resources. They severely oppressed and exterminated enormous portions of indigenous populations either through violent conquest or the exporting of diseases to which native peoples had no resistance. Once established in those areas, they were frequently unable to use those populations for slave labor and therefore imported slaves from other places to plant and harvest food and perform menial tasks. Sometimes the conquered peoples rebelled and expelled the conquerors, creating their own nations and governments. Unfortunately, the temptation to adopt the imperial model, repeating the same patterns that their former colonizers had inflicted on them, proved irresistible.

As Western civilization developed economically as a result of resource conquest, the capitalist system emerged—a system based on ownership of the means of production for the purpose of making a profit. Small business capitalism which was place-based and operated within and for the benefit of a local community morphed into the corporation. Within the past 150 years in America the corporation has generated enormous and frequently obscene profits for a small group of elite owners, and it has also spawned equally obscene abuses of human beings, air, water, land, and animals.

Karl Marx and his devotees have argued that in order in build and maintain a just society, capitalism must be rejected and replaced with a socialistic form of government. In a number of instances, socialist governments have worked well and have been successfully engineered to meet fundamental human needs such as health care, housing, education, the equitable redistribution of wealth, and have implemented a variety of programs to ensure social and economic justice. Not infrequently, however, those governments have deteriorated into authoritarian systems fraught with corruption, and as a result, have become as blatantly oppressive as capitalist systems have been.

Throughout my adult life I have encountered numerous individuals who embrace socialism, and I suppose that if I were forced to choose a modern economic system under which I would prefer living, it would be socialist. However, I have more recently come to understand that merely replacing capitalism with socialism is an ineffectual means of rescuing planet earth and its inhabitants from the cataclysm toward which we seem to be hurtling at locomotive speed. Changing systems is ineffectual not only because an opposite system is still a system, but more importantly because in doing so, only the political and economic structures are altered, while the fundamental paradigm remains largely intact.

During the past eight years I have come to understand that in the United States, unless a candidate is thoroughly subservient to the system, i.e., political and corporate interests, regardless of his or her campaign rhetoric, that individual will not be nominated, let alone elected to the presidency. The same holds true in most cases for nomination and election with respect to lower offices. Furthermore, even the most well-intentioned candidate with the most impeccably progressive voting record and the most principled administration is incapable of addressing, let alone carrying out the monumental changes that are essential in order to prevent collapse. No candidate is humanly capable of executing them in the span of four or eight years, not only because they are so enormous, but because they are impossible to accomplish without a fundamental paradigm shift, not only in the candidate and his or her associates, but throughout the society he or she is governing. And even if such a chief executive could accomplish such changes, they are unsustainable in the current cultural paradigm.

Throughout 2007 we heard a great deal of clamor for impeachment of the President and Vice-President. In 2008 MSNBC released a poll indicating that 89% of Americans favor impeachment. Yet for the same reasons mentioned above, impeachment—which is virtually impossible politically, particularly during a lame duck administration—would have only provided another "bandaid" for the situation. Without a paradigm shift, impeachment would accomplish little in the way of fundamental change because it does not address the root cause of the evil empire's existence and its criminal acts.

Moreover, we now have incontrovertible evidence that America's 2000 and 2004 elections were rigged. Seattle-area grandmother and researcher, Bev Harris, has given us a treasure-trove of documentation regarding election fraud in her marvelous 2006 HBO documentary "Hacking Democracy"[15] More recently, David Earnhardt in "Uncounted: The New Math Of American Elections"[16] has magnificently substantiated the jaw-dropping pervasiveness of election fraud in the United States, how it played out in the elections of 2000, 2004, and 2006, as has author and professor, Marc Crispin Miller in *Loser Take All: Election Fraud and the Subversion of Democracy, 2000-2008*[17].

While it appears that the 2008 presidential election was legitimate, the technology for rigging elections not only exists but has been utilized on several occasions. Since no official investigation of election fraud has ever been carried out in the twenty-first century, the possibility of recurrence is both probable and undeterred.

Geologian, Father Thomas Berry, writes in *The Dream Of The Earth* that, "Whatever their differences, both liberal capitalism and Marxist socialism committed themselves totally to this vision of industrial progress which more than any other single cause has brought about the disintegration that is taking place throughout the entire planet."[18] Indeed, humanity is only beginning to awaken to the vacuousness of the notion of industrial progress, but few of our species are willing to alter our lifestyles in a manner that abandons the old paradigm and embraces a new one.

What *can* alter the paradigm? Tim Bennett asserts in "What A Way To Go" that the culture needs an initiatory experience in order to transform[19]. Harkening back to the myths and stories of the ancients and with the indigenous traditions in mind, Bennett believes that like a young man or woman in a tribal culture, twenty-first century earth dwellers need world-shattering experiences that move the tectonic plates of the psyche and awaken us from the stupor of the paradigm to which we appear to be addicted, thereby causing us to descend into the deeper spiritual layers of our souls. Indigenous initiatory experiences are not merely "rites of passage" but are life-altering encounters with nature and the sacred that permanently mould a young person in a certain fashion which causes her or him to live in a particular manner that he or she otherwise would not have. While the initiation to which Bennett refers has nothing to do with age but rather the nature of one's relationship with the soul, it may produce a fundamental "growing up" or wizening that seems to have eluded individuals steeped in the current infantilizing paradigm.

But what do the elders of indigenous cultures say about initiation? In a 1993 interview with "In Context, A Quarterly of Humane Sustainable Culture", Malidoma Somé of the Dagara tribe of Burkina Faso in West Africa stated that:

> *Growth itself makes one forget about who one is. So initiation is something that is designed to help one remember one's origin and the very purpose of one's occurrence on this side of reality - that is to say, why one was born. This is why initiation is especially magical.*

> *So a person who is not initiated is considered a child, no matter how old that person is, because that person will not be able to recall his or her purpose. Without initiation, the bridge between youth and adulthood can never be crossed, and a person's heart is open to anything - to being shot down by any kind of energy going around. In the village, to not be initiated is to be a non-person.*[20]

Traditional cultures clearly understand that the paradigm of civilization is life-destroying and soul-murdering, having witnessed centuries of brutal devourment of their cultures as a result of it. One of the most, if not *the* most tragic loss in the process has been the destruction of the initiatory experience viewed by the modern technological mind as "quaint," "superstitious", and an "impediment to progress."

Yet the indigenous know that when the initiatory experience is absent or thwarted, when young people are not provided with a structure for initiation by community elders, the initiation will ultimately happen, but rather than occurring in the safety of the sacred tribal container, it will occur in a spontaneously chaotic manner in the world and in distressful life experiences of the non-initiated individual. A conscious initiation by elders results in a grounded deepening of wisdom whereas the absence of initiation leaves the young man or woman untethered and at risk of being battered about by the storms of life. Initiation stamps an indelible mark of blessing on a person's soul; the absence of initiation is a cruel omission that dishonors the soul and leaves it with an insatiable longing which in the current paradigm we attempt to fill with consuming, controlling, and competing.

So when Tim Bennett, "a middle-class white guy" as he calls himself, and Malidoma Somé, a West African tribal elder, tell us that this culture needs an initiation, they are referring to something so profound, so far outside the paradigm, so life-altering that scarcely anything but collapse itself could collectively generate it. The ugly scenarios of collapse that I alluded to above would most certainly be initiatory experiences for many individuals. However, most would not perceive them as such, even though that is precisely what they would be. Unless one comprehends the intention of initiation, it can only feel like torture, injustice, scarcity, terror, and loss, and therefore, the natural tendency in such circumstances is to blame and lash out against one's persecutors. On the other hand, to hold a perspective of collapse as a spiritual initiation does not necessarily alleviate suffering, but it does temper the severity of it with a sense of meaning and purpose.

In *Dreaming The End Of The World: Apocalypse As A Rite of Passage,* Michael Ortiz Hill explores approximately one hundred apocalyptic dreams and echoes Carl Jung's notion that any disturbing thoughts or feelings that we deny or refuse to consciously explore do not vanish, but simply persist, and ultimately, will erupt in our lives in scenarios that are likely to be far more distressing than consciously engaging in the thoughts and feelings we have repressed. Jung emphasized that working with dreams and symbols is often a salutary and less threatening avenue for approaching dreaded psychic material and that, in fact, working with it symbolically may prevent literal manifestations of it in our lives.

Specifically in reference to apocalypse, Ortiz-Hill concludes:

To the exact degree—and this cannot be stated more emphatically—that we do not take apocalypse into the psyche where it truly belongs and suffer through it as a rite of passage, we will be compelled unconsciously to live it out literally to the bitter end.

In a manner that is ultimately mysterious to me, the apocalyptic rite of passage, by consciously bringing to fruition the most difficult realities of the twentieth [or twenty-first] century as they display themselves in one's soul, actually initiates one into the oldest values carried within human culture.

REFLECTION

**What have been some examples of "initiation" in my life?

**How have I dealt with them? What do I wish I might have done differently? What did I do that I'm glad about?

**What are my greatest fears about collapse? How might I be able to lessen those fears?

**What initiatory opportunities might I look forward to in relation to collapse?

NOTES

NOTES

CHAPTER 3–WHAT DOES COLLAPSE REQUIRE US TO FACE?

I find my bearings where I become lost.
~Helene Cixous~

Real safety is your willingness to not run away from yourself.
~Pema Chodron~

Sometimes I find myself unable to wrap my mind around the totality of what the collapse of civilization would actually mean. I know intellectually what losses it would entail, but when I really contemplate them, I feel my eyes glazing and my mind numbing. I have difficulty imagining living without electricity, without heat in a severe winter, without cooling in a torrid summer. Most alarming is the reality that if I don't have enough food stored in my larder, I may not be able to find any in stores, and if I do, the prices may have become unaffordable. But that fear pales when I think about the lack of clean drinking water. Humans can live without food for several days, but not without water.

Families and intentional communities who are consciously preparing for collapse are surrounded by their "tribe", yet how do they navigate the purchase, operation and repair of tools, solar panels, or power generators? Will they have to make their own clothing? If they are employed outside the community or want to be, how long will the employment they might find last? Will Peak Oil have made the transport of food and supplies to local stores impossible? Will hospitals remain open? What will be available in terms of healthcare? How authoritarian will our government become? Will the nation or region be under martial law with National Guard troops and Blackwater Security[21] forces rounding people up and preventing them from remaining in small communities? Or will Peak Oil have made that extremely difficult if not impossible for the powers that will be at that time? Will nuclear war erupt, suddenly extinguishing life on earth, or worse, inflicting radiation sickness and its slow, agonizing death?

I can't answer these questions, but thinking about them is, to put it mildly, alarming. At the very least, issues of fundamental human survival will be paramount. Having enough food, water, energy, and trustworthy human contact will pre-occupy the minds of millions, not to mention the massive displacement of populations and the relocation that some people are already starting to plan for and undertake. Ultimately, what collapse forces us to face is death, so it is not surprising that so few are willing to come to terms with it. (Later sections of this book will specifically connect our fear of death with collapse and hopefully assist the reader in consciously and calmly reflecting on both.)

The magnificent psychiatrist, Elizabeth Kubler-Ross, has given us monumental model for understanding how humans respond to grief, tragedy, and loss. The first stage, **denial**, is when we refuse to accept that this is happening to us. In the case of the collapse of civilization, we've all heard the litanies of denial and probably have a few of our own: Natural phenomena are causing climate change, not humans; there's plenty of oil—all we have to do is drill for it in the United States; the economy is in really bad shape now, but it always bounces back eventually; technology will find a way to fix this—the list is endless.

From denial we then move to **anger**, although frequently, the anger is veiled or misdirected. We may feel angry at our jobs or spouses or children, not fully aware that it is the system of empire with which we feel enraged for its multitudinous violations of our humanity. Or we may feel anger toward the messenger—the doomer/dystopic/downer who announces and articulates collapse in all its fullness. If we are fortunate, we begin to recognize that our anger is at civilization itself, with very good reason, and we focus our anger there, and stop blaming extraneous people and things.

As the reality of collapse bears down upon us, we may then begin to **bargain** with collapse or life or a higher power. With one foot still in the domain of denial, we think to ourselves, "If I drive a hybrid car, maybe I can continue living the lifestyle to which I have become accustomed; if I turn out all the lights in rooms I'm not using, maybe I can postpone the consequences of Peak Oil." We may resort to an endless cache of bargaining chips in order to avoid feeling the full emotional impact of collapse which resides in the next stage.

We can stay stuck in any one of the stages of loss, but genuine healing and empowerment are only possible when we fully open to **grief**—the one emotion that our denial, anger, and bargaining have been defending against. The most compelling and human response to the rape, pillage, and plunder of our planet is deep grief, and paradoxically, it is the gateway to the gifts—yes gifts, that collapse may hold for us.

When we allow ourselves to grieve the innumerable losses which civilization has wrought, our bodies and psyches are freed to move through the grief to a place of **acceptance and re-investment**. As with the other stages, the last is a process, not an event—a process very different from "worry" because what we accept is not only the inevitability of collapse, but our powerlessness to stop it, as well as our limitations in avoiding its daunting repercussions. We accept that there are no guarantees, and as with any terminal illness, we have the opportunity to look fully into the face of our own mortality and beyond. Only then can we wisely re-invest in living our lives, regardless of outcome, which profoundly facilitates responding to collapse with grounded, prudent preparation.

Among those who accept the reality of collapse, debate frequently occurs about whether collapse will occur quickly or over a long period of time. From my perspective, such a debate is irrelevant because collapse will, as Greer has noted, unfold in both ways. Some aspects of collapse will manifest suddenly while others will reveal themselves more subtly over a longer period of time.

In fact, the so-called "slow burn" of collapse didn't just begin with the George W. Bush administration. One case in point is Peak Oil which simply means not the end of oil but the end of *cheap* and abundant oil. For nearly six decades, petroleum geologists and presidents have known about Peak Oil. Yet, at the conclusion of World War II, U.S. auto manufacturers bought up the transit systems of some large cities such as Los Angeles in order to ensure that the overwhelming majority of Americans would have to rely on cars for transportation. No one noticed or cared, really, but today, the consequences of the end of the age of oil for the Los Angeles Basin will be nothing less than catastrophic. From the end of World War II, until very recently, the oil and auto industries have successfully brainwashed the majority of Americans with a preference for the private automobile. It has taken over sixty years for that preference to erode, but with the dizzying fluctuation of gas prices, the likelihood of shortages, or even rationing, and what is at the very least, a brutal economic recession, Americans are beginning to reject the myth of the private car to such an extent that the nation's Big Three automakers are swirling together in a death spiral toward extinction.

I do not consider myself an expert on collapse preparation, nor do I imply that I am qualified to advise readers on relocation. However, it should be obvious that urban areas are likely to be the most distressing venues as food and other shortages, financial meltdown, escalating prices, Peak Oil, climate change, deteriorating infrastructure, epidemic levels of crime and violence, panic, depression, suicide, and myriad manifestations of chaos exacerbate. Some individuals insist that urban areas can be made sustainable and capable of enduring catastrophe with minimal loss of life, but I believe that evidence to the contrary is overwhelming. Neither cities nor suburbs, in my opinion, will be safe or salutary environments as collapse intensifies. Most individuals who take collapse seriously have relocated out of cities and suburbs or are in the process of doing so.

Meanwhile, even as we make our "plans" for relocation and survival, our psyches are brimming with a plethora of emotions. To numerous people, places, and things we must bid farewell. The familiar will be history, and our lives will be about engaging with the unfamiliar, the uncertain, the unanticipated. Loss on the one hand will haunt us, and anxiety regarding the future will be epidemic. We will no doubt reflect on countless myths and archetypes—the Jews wandering in the desert for forty years, Jews many centuries later escaping Nazi Germany in the wake of the impending lockdown of that country, the Hopi in search of the center of the earth. In all of this resides in the back of one's mind, the omnipresent reality that like countless wanderers in humanity's myths, we may not survive, and even if we do, our mortality may claim us long before we consider ourselves "ready" for it to do so. Images of Nazi death camps, torture, firing squads, or permanent separation from loved ones may threaten to engulf us.

In summary, we are likely to be tested emotionally beyond anything we can imagine. And as with most human beings who have come out on the other side of initiatory experiences, we may reflect in hindsight that we don't know how we did it and very often believed we could not. There are no guarantees that any of us will survive; therefore, if we are not prepared to entertain and contemplate the issue of our own mortality, then we cannot realistically prepare for collapse.

Very few individuals in America are willing to even begin contemplating the issues I have just raised, so if you have read this far, I salute you. We are unequivocally in frightening territory. Remember, this is about *meaning,* and finding meaning is arduous work and rarely falls in our laps beautifully gift-wrapped.

You may wonder how I can invite you to consider such dark scenarios as those mentioned above, so it behooves us to take a moment to attend to the notion of "darkness." The poet Rainer Maria Rilke entices us with the promise inherent in darkness:

> *You darkness, that I come from,*
> *I love you more than all the fires that fence in the world,*
> *For the fire makes a circle of light for everyone,*
> *And then no one outside learns of you.*
>
> *But the darkness pulls in everything:*
> *Shapes and fire, animals and myself,*
> *How easily it gathers them—*
> *Powers and people—*
> *And it is possible a great energy is moving near me.*
> *I have faith in nights*

REFLECTION

**What other aspects or repercussions of collapse, not mentioned in this book so far, might I need to face? How do I feel when I contemplate this possibility?

**In terms of Kubler-Ross's stages grief, loss, and tragedy, which stage or stages do I now find myself in? Take some time to journal about your denial, anger, grief, and acceptance/reinvestment.

NOTES

NOTES

Chapter 4–The Darkness and the Divine

I tell you, the day will come when you will review your life and be thankful for every minute of it. Every hurt, every sorrow, every joy, every celebration, every moment of your life will be a treasure to you, for you will see the utter perfection of the design. You will stand back from the weaving and see the tapestry, and you will weep at the beauty of it.

"Friendship With God: An Uncommon Dialogue"
~Neale Donald Walsch~

Western civilization is the product of the heroic attitude depicted in countless myths and fairytales of the past five thousand years. Greek and Roman mythology were replete with tales of the hero's journey—the overcoming of ordeals in order to prove one's faithfulness to the gods and goddesses and one's sense of integrity to the community. The Judeo-Christian tradition further perpetuates heroism in protagonists like Moses, David, Daniel, Jesus, St. Paul, Augustine, the crusaders, and the panoply of saints. The apotheosis of heroics in the Judeo-Christian tradition is the savior who brings salvation. Despite the Enlightenment and the rejection of the mythological, Western civilization has been profoundly and permanently characterized by a heroic attitude. In this country, our Puritan ancestors declared that their fledgling colony was a "city set on a hill", "a light unto the world, "a new Jerusalem"—hence the birth of the American notion of exceptionalism. Like it or not, their work-and-win ethic has permeated our culture, subtly instilling in us the belief that we must survive, conquer, and prevail. "Good" human beings, "morally responsible" Americans want to conquer adversity and win. In fact, to do otherwise implies a deficiency in character.

American exceptionalism has not only inculcated a sense of entitlement, but also an attitude of invincibility. Other nations may collapse, but never the United States because American ingenuity, resolve, and of course, heroism, will find a way to overcome whatever hardships we may face. Technology, modern heroism's preferred handmaiden, will ensure our triumph. Thus, Jared Diamond's notions about the collapse of civilizations have not become wildly or widely embraced in academic or political circles. Nor have Al Gore's "Inconvenient Truth" or Leonardo Di Caprio's "Eleventh Hour" documentaries made a significant impact on the culture, even though they have been viewed by millions around the world, and Gore has won the Nobel Prize for his efforts. In my experience, in most progressive circles, any serious attempt to converse about collapse is viewed as dialog bordering on the fundamentalist Christian rantings of Tim LaHaye and Jerry Jenkins in their "Left Behind" series. Or it is perceived as depressing and pathologically problem-focused.

Heroism on the American left is also manifested in an intractable, and in my opinion, irrational faith in the Democratic and Green Parties to reverse economic, social, cultural, and environmental demise. If the definition of insanity is doing the same things over and over again expecting different results in spite of all evidence to the contrary, then the American left appears to have gone insane. While myriad reasons for this disconnect exist, I believe that heroism is a fundamental factor.

While the culture of Western civilization has been immersed in heroism from its inception, myriad anti-heroic characters abound in its literature such as those of Dionysus, Sisyphus, Holden Caulfield in "Catcher In The Rye", Willy Loman in "Death Of A Salesman"and James Joyce's Leopold Bloom in *Ulysses*. Contrary to the super-human characteristics of the hero, the anti-hero is all-too human—in fact, he or she is very often a miserable failure in many or most aspects of life. But just as most individuals who are products of civilization prefer the heroic, they also cannot tolerate too much of the heroic; we demand characters who exhibit some of the parts of ourselves of which we are ashamed or attempt to conceal from the rest of the world, not merely because we need to know that we are not alone, but because the psyche knows, consciously or unconsciously, that those aspects serve a purpose.

The issue of collapse confronts us not only with scenarios of which we are profoundly frightened, but also with the stark and terrifying reality that our culture, our country, our government, our community, and even our families have not been and will never again be heroic. In fact, collapse unequivocally means that heroics are over. Collapse equals defeat, and the notion of defeat is very, very un-American. And yet, it is precisely our willingness to encounter defeat, despair, hopelessness, powerlessness, loss, and other so-called "negative" emotions which could paradoxically offer "salvation"—not salvation *from* collapse, but salvation *from continuing to deny that collapse is happening all around us*—and salvation from the toxic legacy of empire. Our willingness to feel the "dark" emotions could open the door to authentic empowerment, or as Pema Chodron states, *Real safety is your willingness to not run away from yourself.*

Denial requires an enormous amount of energy, but once that energy is no longer being consumed in blockading the door lest the "collapse monster" enter, we are energetically freed up to begin the work of directly engaging the monster. As Jung said, what we deny does not depart, but only further develop.

Moreover, were Jung with us today he would undoubtedly diagnose America as having been infected with mass psychosis, and his prescription would be that we face our darkness deliberately. At this moment in history the fabric of America is reverberating with financial tremors set off by a bursting housing bubble, massive foreclosures and bankruptcies, skyrocketing food prices, untold masses of individuals drowning in debt yet relying on credit cards for basic necessities, monthly escalating unemployment statistics, a stock market on life support, a federal government bailing out insolvent corporations in tandem with a Federal Reserve gone berserk printing money out of thin air. The U.S. financial system is in tatters, writhing in full-blown collapse. One aspect of our mass psychosis has been consumerism—buy now, pay later; spend what we don't have, flip that house— refinance, put it on plastic, we deserve it, this is America for god sake! And now the bills—oh so many different kinds, on so many different levels, are coming due, and it's time to pay the piper.

Jung would remind us of archetypes of the soul—those universal themes or motifs found in literature, art, music, story, and deep within the human psyche. He would also emphasize that each one has a polar opposite. Since the end of World War II, America has been deliriously reveling in abundance, the polar opposite of which is scarcity. Never again, we told ourselves, would we have to experience the deprivation of a Great

Depression. We can have it all—or so we thought, but the scarcity monster that we refused to look in the eyes—reflected in the faces of millions of hungry humans worldwide, off of whose labor and resources we were able to feast—that monster has now shattered the door behind which we locked him, and he is here, threatening to include our own faces among the hungry and exploited while American school children load up their backpacks with food on Fridays to feed their entire families over the weekend and food banks struggle from day to day to feed the hungry. Yes, it's happening now—in America.[22] When darkness is not dealt with, it only grows darker and more insistent on being acknowledged.

But economic collapse is only one facet of the darkness that is bearing down upon us. I've already mentioned a litany of others. There's nowhere to hide; it's global now, engulfing the planet and every living being on it. Setting this book down and calling it "doomerish" doesn't change the fact that we are currently witnessing the extinction of 200 species per day on planet earth. It won't reverse Peak Oil or climate change or global famine or endless resource wars. Out of civilization have evolved monsters that refuse to be silenced, refuse to disappear, and refuse to allow business as usual to continue for one more moment.

If the heroic is over, then what is the alternative? In a word, *surrender*. Surrender does not mean weakness, capitulation, lack of courage, or poor character. What it does mean is acknowledging our defeat and allowing that to open the door to the opportunities waiting to reveal themselves when we stop denying reality.

No one could have stated it more beautifully than Kahlil Gibran in his poem "Defeat" from "Madman" in 1918:

Defeat, my Defeat,
my solitude and my aloofness.
You are dearer to me
than a thousand triumphs,
and sweeter to my heart
than all world glory.

Defeat, my Defeat,
my self-knowledge and my defiance.
Through you I know that
I am yet young and swift of foot
and not to be trapped
by withering laurels.
And in you
I have found aloneness
and the joy of
being shunned and scorned.

Defeat, my Defeat,
my shining sword and shield.
In your eyes I have read
that to be enthroned
is to be enslaved,
and to be understood
is to be leveled down,
and to be grasped
is but to reach one's fullness

and like a ripe fruit
to fall and be consumed.

Defeat, my Defeat,
my bold companion,
you shall hear my songs
and my cries and my silences,
and none but you shall speak to me
of the beating of wings,
and urging of seas,
and of mountains
that burn in the night,
and you alone shall climb
my steep and rocky soul.

Defeat, my Defeat,
my deathless courage,
you and I shall laugh together
with the storm,
and together we shall dig graves
for all that die in us,
and we shall stand
in the sun with a will,
and we shall be dangerous.

Bill Plotkin, author of *Nature And The Human Soul*[23], states that a mature ego "understands the occasional necessity of surrendering to or being defeated by a force greater than itself" because our power is not in heroism, but in surrender: "We shall stand in the sun with a will, and we shall be dangerous," says Gibran. Willingness to encounter the darkness rather than running from it is empowering in ways we cannot imagine because within that darkness resides the divine. As Jung often noted, gold is not found in lovely green meadows bathed in resplendent sunlight. It is found in the depths of the earth—in locations dark, dirty, treacherous, and difficult to navigate. And even after being extracted from the earth, gold like the diamond, does not glitter until it has been mined and polished. Jung would tell us in 2009 that if we are willing to descend into the darkness, we will find the gold, but it will be a daunting and messy task. I cannot help but find irony in this since many individuals preparing for collapse who believe that the currencies of many nations may become worthless in the process have been buying gold for some time. Thus, it seems that "mining the gold" is an apt image for our time.

REFLECTION

**In what ways have I attempted to be heroic in my life? What were the results?

**In what ways have I been anti-heroic, that is, sometimes less than perfect, incompetent, under-achieving, or not living up to this culture's standard of unabated success? What in my anti-heroism has been valuable, perhaps even appealing, for me?

**In what ways have I been able to surrender to defeat? What was the outcome?

**What "gold" have I discovered in my own personal adversities?

**In what ways might I still be in denial about the reality of collapse? What prevents me from rejecting denial and facing what is most difficult to face?

**What has been the result of my moving through denial about collapse?

NOTES

NOTES

Chapter 5—A Culture of Two Year-Olds and the Gift of Collapse

The American way of life is not negotiable.
~Vice-President Dick Cheney~

The problem with this statement is that if you are unwilling to negotiate, you get a new negotiating partner called reality.
~James Howard Kunstler~ in
"Escape From Suburbia"

The only certainty regarding collapse is all of the uncertainty it holds. Will it be fast or slow? Food, water, learning survival skills, relocation? One feels almost overwhelmed just thinking about all of the issues contained in the collapse of civilization, but perhaps none is as formidable as being forced to live very differently than we do now. Energy depletion and global economic crisis guarantee that the resources that fuel our current American middle-class lifestyle will not be available. Peak Oil means not only wildly fluctuating gas prices, but skyrocketing prices for food and many other items that we now take for granted. The Great Depression generation experienced shortages, and so will we, but Americans are unable to imagine the extent to which collapse will curtail their lifestyles. In fact, when I've raised these issues in classes or groups to which I've spoken, the responses have sometimes implied that the notion of having to limit one's consumption to such a severe extent is nothing less than "un-American", echoing once again American exceptionalism and entitlement. My other favorite response of course, is that I'm hopelessly negative and short-sighted because technology will always find a way out.

As an historian I can freely use *exceptionalism* and *entitlement* to describe the attitudes above, but I was not always an historian. My training in psychology causes me to look into the deeper layers of an almost tantruming insistence that as Americans we "shouldn't have to" do without or be deprived. It's our inherent "right" to own SUV's, plasma TV's and 6000 square-foot houses. Yet even those touting these "rights" know in some part of their psyches that their argument is irrational. If the majority of the earth's inhabitants cannot afford these things, then how can it be a "human right" or even an "American right" to insist on having them?

As Richard Heinberg states in "What A Way To Go", our culture has infantilized us to such an extent that we illogically assume and even demand that we must have our gas-guzzling vehicles and all of our other "toys" that make an enormous energy footprint on the earth. And in the same documentary, its producer Sally Erickson, says it even more succinctly when she states that it is as if we are a culture of two year-olds who refuse to accept limits. Civilization's refusal to accept limits has made humans not only self-absorbed, greedy, grasping, and "entitled", but it has also prevented them from developing the maturity to commit to leaving a sane and secure world to their children.

Few of us remember ourselves at the age of two, but most parents are quite familiar with two year-old behavior. The two year-old wants what she wants when she wants it, and her favorite word is "No!"—as long as she can *say* "No!" and doesn't have to be *told* "No!" At the age of two, children want to believe that they are omnipotent and have absolutely no limits. Their developmental task, however, is to be able to say "No" but also be able to accept the finality of being told "No". One of the reasons parents and psychologists refer to the two's as "terrible" is that caretakers must walk a very fine line between allowing the child the right to say "No" but at the same time setting limits. The second year of life is a time to begin making choices but choices based on the limits that are set by caretakers.

The infantalization technique of civilization is obvious in its ability to "hook" the two year-old in us, deluding that part of us into believing that there are no limits and that we can have whatever we want—and not only that we can have it, but that we *should*. We prefer to be omnipotent with our money, our time, our environment, our relationships, and with most other aspects of life, but ultimately, collapse will not permit us to be.

For this reason, when fully in the throes of collapse many individuals will feel "abused", "battered", "persecuted", and "victimized", and they will not experience the limits that collapse is setting on them as in their best interests. In fact, like children, they will look around for someone else to blame. Thus, those who understand collapse— why it is happening and that it *is* happening and who because they comprehend the finality of it have been consciously preparing for it, will feel less victimized and will have more energy to actually participate positively in the experience of collapse.

While collapse brings hardship and sacrifice, it also brings opportunity for those who are awake to it to swim in its river of transition. In so doing, the omnipotent two year-old who formerly refused to accept limits has the option to grow up and become an adult who makes conscious choices about what will serve him/herself, the community, and the earth.

Even as I have been called a "doomer" by those who do not understand the full spectrum of collapse, I have continued to insist that the news is not all bad. I see little indication from individuals on the progressive left, and almost no indication from individuals on the conservative right or those in the center, that they are willing to accept the limits that will be foisted on them by collapse. In fact, the notion that their lifestyles will be constrained by anything would strike them as profoundly absurd. Yet those who understand collapse also understand that such constraints are inherent in the process.

Like an exasperated parent or a stern elder, collapse will force unprecedented limits on the human race and compel a descent into the underworld of radical initiation described above. And as with all initiations, the outcome is unknown. There is never a guarantee that young women or men in tribal initiations will prevail instead of perish. While I do not rule out the possibility of a mass transformation of consciousness, I see no evidence of it at this moment but rather, monumental evidence to the contrary. So far, it appears that the only way in which humanity can be initiated and reborn is through widespread and catastrophic collapse.

Yet even as I write of this frightening reality, my heart swells with excitement as I think of the individuals who are reading the signals, learning skills, relocating to more sustainable environments, and making the sacrifices collapse is demanding of our species. I imagine those who will, with reverence for the earth, grow glorious gardens; compost and recycle everything they possibly can; build beautifully-designed natural habitats; make their own clothing and furniture; home school their children; build extraordinary, vibrant communities that serve and support one another; utilize herbal and other natural healing techniques; discover the joys of leisure spent with each other in a world where a power grid no longer exists, as people laugh, love, talk, make music, dance, tell stories, play, and commune with each other. Yes, perhaps all of this will occur against the backdrop of pandemics, natural disasters, climate change, or even a nuclear exchange, and certainly not everyone will survive. Yet I am unwilling to forego the rebirth aspects of initiation because they are always potentially present, without exception. As stated above, I must hold the vision of rebirth alongside current reality in order to have an accurate picture of the whole.

In a world where the fear of endings seethes just under the surface of rampant consumerism and in our DNA and while fundamentalist Christians prognosticate about the "Second Coming", William Butler Yeats' poem by that title heralds the rebirth implicit in civilization's demise:

Turning and turning in the widening gyre
The falcon cannot hear the falconer;
Things fall apart; the centre cannot hold;
Mere anarchy is loosed upon the world,
The blood-dimmed tide is loosed, and everywhere
The ceremony of innocence is drowned;
The best lack all conviction, while the worst
Are full of passionate intensity.

Surely some revelation is at hand;
Surely the Second Coming is at hand.
The Second Coming! Hardly are those words out
When a vast image out of Spiritus Mundi
Troubles my sight: somewhere in sands of the desert

A shape with lion body and the head of a man,
A gaze blank and pitiless as the sun,
Is moving its slow thighs, while all about it
Reel shadows of the indignant desert birds.
The darkness drops again; but now I know
That twenty centuries of stony sleep
Were vexed to nightmare by a rocking cradle,
And what rough beast, its hour come round at last,
Slouches towards Bethlehem to be born?

Collapse is a form of death, and Americans do not like the word "death." We go to extraordinary lengths to dress it up, pretty-fy it, deny it, and as my favorite of all meaningless anti-death cliches goes, "put it behind us." Like banshees, we drive ourselves heroically in the first half of life as if there were no death. It will engulf others but not us. Remember, we are the "exception,"; others will die, not us. Other civilizations will collapse; not ours. Yet it was Jung who said that, "There is a great obligation laid upon the American people—that it shall face itself—that it shall admit its moment of tragedy in the present—admit that it has a great future only if it has the courage to face itself."[24] America the nation is not likely to "face itself," but as individual Americans, we must, if we intend to successfully navigate collapse.

I too resist collapse, but at the same time that I resist it for similar or different reasons from those around me, I am also consciously working to embrace it. To embrace something or someone is not necessarily to throw one's arms wildly around that event or person, but to slowly, intentionally open to the gifts inherent in what we most dread. I do not say this lightly. I am a survivor of breast cancer. My world "collapsed" fifteen years ago when I was diagnosed with it. But as is frequently the case, my world was also transformed by a terminal illness, and I became a different person as a result of it. Again, Pema Chödrön writes, "Openness doesn't come from resisting our fears, but from getting to know them well."[25]

So what is the "gift" of collapse, or more accurately, the "gifts"?

First, collapse strips us of who we think we are so that who we really are may be clearly revealed. Civilization's toxicity has fostered the illusion that one is, for example, a professional person with money in the bank, a secure mortgage, a good credit rating, a healthy body and mind, raising healthy children who will grow up to become successful like oneself, and that when one retires, one will be well-taken-care of. If that has become our identity, and if we don't look deeper, we won't discover who we really are, and as collapse intensifies, we will be shattered because we have failed to notice the strengths, resources, and gifts that abide in our essence which transcend and supersede one's ego-identity. In a post-collapse world, academic degrees and stock portfolios matter little. The real question, as Richard Heinberg so cogently asks, is: Do you know how to make shoes?

Just ask countless individuals who have had everything stripped away as a result of speaking truth to power. One day they were "solid citizens" with sterling careers; the next day, they were "enemies of the state" fearing for their very lives. We can learn much from their journeys about preparing for life after collapse.

In *Nature And The Human Soul,* Plotkin has provided us with a look at how differently humans might develop psychologically in an eco-centered, rather than an ego-centered society. An ego-centered society serves the needs of industrial civilization and the consumption of its products whereas in an eco-centered society, "its customs, traditions, and practices are rooted in an awareness of radical interdependence with all beings." Thus, the crux of the initiatory experience is the transformation of the childhood or adolescent ego into a fully developed, fully human individual who grasps wholly her interdependence and recognizes her purpose, that is, what she has come here to do.

One way to prepare for the initiatory experience of collapse is to explore the issue of identity apart from one's social roles. An individual may identify as an accountant, a teacher, an engineer, a therapist, but only in moments of solitude, away from the daily role of our profession, do we have the opportunity to explore who we really are. For me, a spiritual path has been crucial in assessing who I am apart from what I do, and much more will be offered in subsequent chapters on this topic.

Secondly, collapse will decimate our anti-tribal, individualistic, Anglo-American programming by forcing us to join with others for survival. You may own a home outright with ample acreage on which you have produced a stunning organic garden, have a ten-year cache of food and water, drive a hybrid car, and live a completely solarized life, but if you think you will survive in isolation, you are tragically deluded. Collapse dictates that we will depend on each other, or we will die.

I have been an activist for over thirty years. Without exception, every time I have been involved with other activists in promoting change, personalities clash, egos become bruised, people tantrum, become disillusioned, and walk away from the group. We all seem to have Ph.D.'s in "self-sufficiency" but remain tragically ignorant of genuine cooperation. We will transform this pattern as civilization collapses, or we will perish, and the process of that transformation probably won't be a pretty picture. However, we can begin preparing in present time for the collective thinking and action that collapse will necessitate by, for example, developing small dialog circles of community[26] which meet regularly to engage in deep listening and truth-telling about how we are experiencing collapse. (Chapter 12 will address dialog circles in depth and offer instructions for facilitating and participating in them.)

Not only will we be compelled to relate differently to humans, but to all beings in the non-human world as well. Only as we begin to read the survival manuals that trees, stars, insects, and birds have written for us, will our species be spared. The very "pests" that we resent as unhygienic or annoying may, in fact, save our lives. One year ago, the honey bees used to circle around me on warm days when I ate my lunch outside under the trees, sitting on the grass. Today, I sit under the same trees on the same grass, but the honey bees are gone. No one seems to be able to tell us why. Maybe it's time to ask the bees to tell us why.

Paradoxically, collapse may bring meaning and purpose to our lives which might otherwise have eluded us. In our linear, progress-based existence, we rarely contemplate words like "purpose." With civilization's collapse, we may be forced to evaluate daily, perhaps moment to moment, why we are here, if we want to remain here, if life is worth living, if there is something greater than ourselves for which we are willing to remain alive and to which we choose to contribute energy. These decisions probably will not be made in the cozy comfort of our homes, but in the streets, the fields, the deserts, the forests, in the eerie echoing of our voices throughout abandoned suburbs, and beside forgotten rivers and trails. Purpose will rapidly cease being about what we can accomplish and will increasingly become more about who we are. In a collapsing world, the so-called "purpose-driven life" will no longer exist. Humans will be "driven" by the determination to survive and assist loved ones in surviving and to plant the seeds of life lived in intimacy with the earth community. From that quest for survival and transformation will emerge authentic purpose, which will undoubtedly not resemble anything we can imagine today.

Lest the reader infer that I'm portraying collapse as some exercise in airy-fairy spirituality devoid of practicalities, I hasten to re-emphasize what I have stated above-- that collapse will require humans to attend to the most pragmatic realities of existence—food, water, shelter, health care, and a host of other survival issues. As centralized systems such as federal, state, and local governments are eviscerated, communities will be compelled to unite in order to solve these issues—to grow gardens, make clothing and other items, treat each others' illnesses, birth and bury one another, create community currencies, and rebuild infrastructures on an intensely local level.

A poem by one of America's most under-rated poets, William Stafford, assures us that that the fears we take into ourselves will bless us. In fact, the poem "To My Young Friends Who Are Afraid", suggests that we find our way *by* being afraid.

There is a country to cross you will
find in the corner of your eye, in
the quick slip of your foot—air far
down, a snap that might have caught.

And maybe for you, for me, a high, passing
voice that finds its way by being
afraid. That country is there, for us,
carried as it is crossed.

What you fear
will not go away: it will take you into
yourself and bless you and keep you.
That's the world, and we all live there.

The quality of spirituality that may emerge from attending to the fundamentals may be an authentic "fundamentalism" in the truest sense of the word. In a post-collapse world, "fundamental" spirituality will be about caring for the basic needs of loved ones, becoming nurturing stewards of the ecosystem in whatever condition it may be at that time, noticing what one now values as opposed to what was most important prior to collapse— seeing, hearing, smelling, tasting, feeling all aspects of existence to which we were oblivious, or only mildly attentive, before the distractions were stripped away. Certainly, this is not likely to be the comfortable, privileged, indulgent spirituality of the New Age workshop circuit, but may more closely resemble the earth-based reverence for the sacred that our tribal ancestors so dearly cherished and demonstrated.

Spiritually, we can now begin preparing for the collapse of civilization as we have known it by opening ourselves each day to the "lesser collapses" of civilization that we see around us, such as the loss of a viable, uncorrupted electoral process, the demise of centralized systems and corporations that no one ever thought would go bankrupt, the decay of infrastructure, and the deterioration of institutions such as education, religion, health care, and the legal system. Human beings have had several thousand years to create functional societies, and in many cases, they have. Those civilizations have also collapsed because all civilizations ultimately do. The United States has had 233 years to fashion a sustainable nation. With the death of Abraham Lincoln at the end of the Civil War, corporations and centralized systems triumphed in controlling every aspect of American life, and they have been doing so until the present moment. Thus, not surprisingly, in the 1970s when corporate America knew very well that U.S. oil production had peaked and that within three decades, the nation and the world would be confronting a catastrophic energy crisis, it did absolutely nothing, choosing rather inebriation with the profits of hydrocarbon energy and the suppression of alternative technology rather than assisting the nation in building lifeboats for navigating a post-petroleum world.

For millennia, many indigenous people have described the demise of civilization we are now witnessing as a purification process—a time of rebirth and transformation. Their ancient wisdom challenges us to face with equanimity the collapse that is in process; that is, to hold as much as humanly possible in our hearts and minds the reality of the pain the collapse will entail, alongside the unimaginable opportunities it offers. Or as Pema Chödrön would say, "Get to know collapse well."

Some people tell me that they would rather not know what's going on because they prefer to live their lives from day to day doing the best they can to make a better world, enjoying their loved ones, and earning their bread. I certainly understand their desire to protect themselves from the pain of awareness, but I also know that they are exchanging long-term preparedness for temporary comfort and that the pain of awareness in present time is far less than the pain they will incur as a result of ignoring it.

As I have stated above, none of us is facing collapse by accident. Our challenge is to open to it with the awareness that it will bring gifts alongside adversity. Or to quote Pema Chodron again, "Only with this kind of equanimity can we realize that no matter what comes along, we're always standing in the middle of sacred space. Only with equanimity can we see that everything that comes into our circle has come to teach us what we need to know."

REFLECTION

**What are some of the lesser collapses to which I have opened in my life? Are there others that I would like to explore and continue opening to?

**What are some limits I need to set on my inner two year-old around preparation for collapse? What are my fears around setting such limits? What limits have I already set? What has been the result of setting these limits?

**What are some ways that my adult self is preparing or has been preparing logistically for collapse? What have I noticed and how have I felt in making those preparations?

**What is my "spiritual path"? What does that mean to me? Take time to journal about this. If you don't have a spiritual path, where do you find meaning and purpose in your life? Are those avenues of meaning and purpose useful to you in preparing for collapse? If so, elaborate in your journaling. If not, what else might you need to do to enhance meaning and purpose in your life?

NOTES

Notes

Chapter 6–Who Are You Really, Wanderer?

A Story That Could Be True

If you were exchanged in the cradle and
your real mother died
without ever telling the story
then no one knows your name,
and somewhere in the world
your father is lost and needs you
but you are far away.

He can never find
how true you are, how ready.
When the great wind comes
and the robberies of the rain
you stand on the corner shivering.
The people who go by--
you wonder at their calm.

They miss the whisper that runs
any day in your mind,
"Who are you really, wanderer?"--
and the answer you have to give
no matter how dark and cold
the world around you is:
"Maybe I'm a king."

~William Stafford~

On December 17, 2008, a Reuters story "Downturn Spurs Survival Panic" reported that, "A paralegal, recently laid off, wanted to get back at the 'establishment' that he felt was to blame for his lost job. So when he craved an expensive new tie, he went out and stole one.

The story, relayed by psychiatrist Timothy Fong at the UCLA Neuropsychiatric Institute and Hospital, is an example of the rash behaviors exhibited by more Americans as a recession undermines a lifestyle built on spending."

In the coming months, the story continues, "mental health experts expect a rise in theft, depression, drug use, anxiety and even violence as consumers confront a harsh new reality and must live within diminished means."

In yet another 2008 story, "The Great Accumulation Hits The Wall", the *Wall Street Journal* reported that:

> *On Black Friday, the day after Thanksgiving and the first official day of the holiday shopping season, 31-year-old confessed shopaholic Nikki Ebben was holed up in her bedroom in Appleton, Wis., while her husband went to Wal-Mart to snag a $500 flat-screen TV. Ms. Ebben, who has maxed out 15 credit cards and racked up more than $80,000 in debt, says she vowed to stay away from stores. Still, she couldn't resist the temptation of e-commerce, particularly the appeal of 30% off and free shipping. While her*

husband was gone, she spent $400 at Toysrus. com and Target.com, using money from the couple's joint bank account.

"I went crazy", admits Ms. Ebben, whose mother stopped speaking to her for a time because she owed her parents so much money.

Both the Reuters and *Wall Street Journal* stories conclude that buying and consuming have become part of the national culture, offering people an identity—the identity of a consumer, which many will now be forced to abandon. Additionally, shopping has become a way for countless individuals to cope with their emotions. Not only do the things we buy allow us to feel good momentarily, but the disease of consumerism has become so pathological that in many instances, people have come to believe that they *are* what they buy, and the more expensive and coveted brand or product makes a statement about who one is. This is enormously significant because there's obviously more than "survival panic" going on here.

I believe that it's not a stretch to conclude that for some, the inability to consume may be creating a fundamental existential crisis in terms of losing one's identity. This would certainly explain the bizarre violence that occurred at the Long Island Walmart on the day after Thanksgiving, 2008, where an employee was trampled to death. If consumption "rewards" human beings with a positive identity as well as the sense of financial security, then it is nothing less than an extremely powerful addiction. Withdraw the addictive substance or activity—or put it on sale at 70% off, and many people will behave like the street junkie who will do whatever it takes to score his next fix.

I hasten to add that short of living on air, none of us can totally cease consuming. The issue, of course, is not consumption itself, but consumption that isn't about buying or bartering for what we truly need—consumption based on fear, insecurity, alienation, all of which are rooted in the human ego, as opposed to the human soul.

While we may want to shake our heads when hearing reports like the ones above, I consider them very positive aspects of economic collapse. Yes, these desperate individuals are suffering a plethora of emotions as they are forced into withdrawal from their drug of choice, shopping, but in my opinion, this *is* the upside of the unraveling. There are no guarantees that any of them will experience a personal psychological or spiritual epiphany regarding the meaning of life, but as with any addiction, when the drug of choice is no longer available, an opening exists for the addict to make a different choice which may not have been possible without forced withdrawal. Welcome to cultural rehab in the throes of the collapse of Western civilization!

In a culture where tribal community, intimate connection with nature, and concomitant initiatory rites are absent, then the human psyche, which appears to inherently require these for optimum functioning, will consciously or unconsciously devise its own rituals for constructing an identity. If this is so, then we may conclude two things: that initiation makes mindless consumption unnecessary, and that mindless consumption in search of identity is a substitute for initiation.

In indigenous/traditional cultures, even when entire nations or villages are steeped in poverty, there is almost always a curious sense of "enough." In fact, one usually finds there, more generosity, magnanimity, and compassion than anywhere in industrial civilization. One reason for this may be that beyond a sense of "I *have* enough because the tribe shares with me and I with them" is a more fundamental sense of "I *am* enough because I know who I am."

As mentioned numerous times above, the most important issue—the pivotal question as we attempt to navigate collapse is: *Who am I?* As I've also been saying, collapse is likely to shatter our previously-held identities and open us to an "identity crisis" such as we have never known. William Stafford implies that the "great wind" and the "robberies of the rain" that cause us to "stand on the corner shivering" may be more bearable if we have come to understand that in our core, we are a king or a queen.

Why is the question of *Who am I?* so momentous? Isn't focusing on this question another form of navel-gazing self-absorption? How can I be preoccupied with philosophical questions in the midst of everything else I will have to attend to in order to navigate collapse? Shouldn't I be just as concerned with other suffering souls around me as I am with who I am?

One of the first principles we must establish as we consciously face collapse is that polarizing opposites do not serve us in times of initiation. So a deeper reflection on each of the above questions in the light of paradox, that is, apparent contradiction that actually is not, reveals that the proper attitude in attempting to answer them is not either/or, but rather, both/and.

Reflecting on *Who am I?* can turn into navel-gazing, but it does not have to. And yes indeed, we must all be concerned with the suffering of those around us, but the *Who am I?* question does not preclude that concern. Just as we are ultimately "kings" and "queens" in our essence, so is every other earthling. Furthermore, that question does not have to be an additional burden, but rather, a liberating force, in the midst of all that we must attend to in navigating collapse. In fact, it may not be possible to navigate it *unless* we continuously reflect on who we are.

Sooner or later collapse will force humans to interact directly with nature. It may happen as people intentionally relocate and plant and harvest food, catch rainwater, butcher animals for food, discern seasonal changes in relation to survival needs, and in hundreds of other ways. For those who may end up homeless and on the streets as a result of compounded losses, nature will exert a variety of hardships. Many individuals believe that it is in their best interests to consciously begin developing their relationship with nature now, rather than when they are forced to do so. I strongly resonate with the latter group and have for decades, having discovered that the crux of my spiritual path lies in my connection with nature.

Author and psychotherapist, Thomas Moore writes that "...nature gives us the most fundamental opening to spirit."[27]

If we recall our hunter-gatherer ancestors described above, we realize that they held a deeply intimate relationship with nature; in fact, their lives depended on that relationship. Today, we live in civilized societies that dominate nature, and we have been taught that we need not "bother" with communing with it. However, collapse will force the cessation of institutionalized domination of nature, and few will survive who are not in relationship with it. Moreover, it is within that web of connectedness with nature that we are best positioned to answer the question of *Who am I?*

The Lakota Sioux elder, John Fire Lame Deer famously stated:

> *Let's sit down here...on the open prairie, where we can't see a highway or a fence. Let's have no blankets to sit on, but feel the ground with our bodies, the earth, the yielding shrubs. Let's have grass for a mattress, experiencing its sharpness and its softness. Let us become like stone, plants, and trees. Let us be animals, think and feel like animals. Listen to the air. You can hear it, feel it, smell it, taste it. Woniya Wakan—the holy air—which renews all by its breath. Woniya, woniya wakan—spirit, life, breath, renewal—it means all that. Woniya, we sit together, don't touch, but something is there; we feel it between us, as a presence. A good way to start thinking about nature, talk about it. Rather, talk to it, talk to the rivers, to the lakes, to the winds as to our relatives.*

Our indigenous ancestors have revealed unequivocally that they could not survive without a deeply personal connection with nature. The Lakota gave us the beautiful expression *Mitakuye Oyasin* or "all my relations"—meaning that we are related to every member of the non-human as well as human world. Native peoples often speak of "standing people"(trees), "fish people", or "stone people" as if trees, fish, and rocks are persons to be communed with, not objects to be possessed.

Thomas Moore underscores this notion:

> *Nature is not only a source of spirit; it also has a soul. Spiritually, nature directs our attention toward eternity, but at the same time it contains us and creates an intimacy with our own personal lives that nurtures the soul. The individuality of a tree or rock or pool of water is another sign of nature's soul.*[28]

In an article "The Psychological Benefits Of Wilderness", Garrett Duncan of Humboldt State University surveys a number of studies which document the therapeutic effects of wilderness experiences.[29] Likewise, Theodore Roszak in his article "Awakening The Ecological Unconscious"[30] states that "...repression of the ecological unconscious is the deepest root of collusive madness in industrial society; open access to the ecological unconscious is the path to sanity."

No matter how estranged we may feel from nature, something in our ancient memory recalls our intimacy with it. We may know little about organic gardening or composting, yet that memory lives. Therefore, everything we need to restore our connection with nature is already available to us. The best place to start is not necessarily in planning next year's garden now. If our lifestyle has been one of alienation from nature, then the easiest way to begin reunion with it might be to simply take ourselves out into it for an entire day—walking, lying, sitting, meandering in a woods, in a meadow, along the shore of a river, lake, or ocean. A useful tool is a journal where we might do some writing or sketching of images. (A section at the end of this chapter has been provided for journaling.) Above all, we should quiet ourselves and simply listen, leaving the Ipod at home and listening instead to the sounds, smelling the smells, seeing the colors, shapes, textures, light, and shadows of nature.

During this time it is important to notice our resistance to experiencing nature—the anxiety, fidgetiness, boredom, emptiness, loneliness, fear, or other "negative" emotions that may arise. Notice what irritates, frustrates, saddens, or incites fear. Also, notice what is lovely, engaging, beautiful— what feels nurturing, comforting, joyful, exciting, deliciously wild and untamed. Journaling or drawing these emotions may be useful.

The words of ecologist and author of *The Spell Of The Sensuous*[31], David Abrams further illuminate our potential for intimacy with nature:

> *There is an intimate reciprocity to the senses; as we touch the bark of a tree, we feel the tree touching us; as we lend our ears to the local sounds and ally our nose to the seasonal scents, the terrain gradually tunes us in, in turn. The senses, that is, are the primary way that the earth has of informing our thoughts and of guiding our actions. Huge centralized programs, global*

initiatives, and other 'top down' solutions will never suffice to restore and protect the health of the animate earth. For it is only at the scale of our direct, sensory interactions with the land around us that we can appropriately notice and respond to the immediate needs of the living world.

Equally useful will be our experience of re-entering civilization. What feels abrasive, dissonant, assaulting, and alienating? If we experience "relief" in returning to civilization, it's important to notice why. In summary, it is important to notice and feel the contrast between being present in nature and returning to our civilized existence.

For those not yet intimately connected with nature, I suggest that this experience be repeated several times, retreating to nature and returning to civilization. The contrast needs to be deeply felt.

For those individuals already quite comfortable in nature and its solitude, then the next step is to begin the journey of learning specific skills that will be essential for navigating collapse. I'm talking not only about organic gardening, but becoming familiar with wild plants and herbs, raising poultry, milking cows, butchering, hearth cooking, canning, food drying techniques, woodworking, weaving, emergency first aid, healing with herbs and homeopathic substances, and other survival skills.

Carolyn Baker, Ph.D.

But along with the *doing* that we must all attend to in preparation for collapse, it is crucial that we do not lose our *being*. Even as we strategize regarding relocation, perhaps even consider expatriating, and learn the survival skills necessary for navigating collapse, we must continue to ponder "Who am I in the face of collapse? Who am I as the world as I have known it ends?"

Who is any of us apart from our human parents? Indeed we are far more than the children of our human parents, and that is so, in part, because we are also children of the earth.

Father Earth

There's a two million year-old man no one knows.
They cut into his rivers.
They peeled wide pieces of his hide from his legs.
They left scorch marks on his buttocks.
He did not cry out.
No matter what they did, he did not cry out.
He held firm.
Now he raises his stabbed hands and whispers that
*   we can heal him yet.*
We begin the bandages. The rolls of gauze. The gut,
*   the needle, the grafts.*
We slowly, carefully, turn his body face up.
And under him, his lifelong lover, the old woman is
*   perfect and unmarked.*
He has laid upon his two million year-old woman
*   all this time*
Protecting her with his old back, his old scarred
*   back.*

84

> *And the soil beneath her is fertile and black with her tears.*
> ~Clarissa Pinkola Estes~

Some years ago, I heard Clarissa recite this poem in a workshop in which she told the story of Ishi, the last Yahi Indian who lost his entire family and tribe and was discovered by whites in Oroville, California in 1911. When Ishi was found, he was terrified and disoriented after walking out of the Sierra Madre Mountains, not knowing where to go or what to do. Eventually, he was transported to the University of California anthropology museum in San Francisco, where a compassionate anthropologist, Alfred Kroeber, befriended Ishi, and with his staff, documented the Yahi language and gleaned a treasure-trove of information about Yahi culture.

Although Kroeber considered Ishi his friend, many of his colleagues perceived Ishi as an "object" to be studied rather than a fellow human being with whom they might develop a relationship. The ultimate objectification occurred in 1916 when Ishi died while Kroeber was traveling in Europe. In Europe Kroeber received a letter from the museum staff stating that Ishi had become ill with tuberculosis, and the prognosis was not favorable. Kroeber immediately sent a letter to the staff unequivocally forbidding them to perform an autopsy on Ishi. Sadly, the letter arrived too late, and Ishi died before Kroeber returned, and yes, not only was an autopsy performed, but Ishi's brain was removed and sent to the Smithsonian. It was later recovered and sent to Ishi's closest living relatives in California.

Many of us living in the culture of empire may feel like Ishi from time to time—as if we are from another world, another time, and as if we have lost everything that is truly natural and innate to us. Certainly, civilization presents many opportunities for us to feel objectified and estranged from nature and the sacred. While none of us has lost what Ishi lost, the collapse of civilization may take us very near the edge of loss as abject as Ishi's where we have nothing to which we can cling except "Father Earth and the woman who lies beneath him".

Collapse may not remove everything familiar, but it will strip us of so much that presently provides identity and grounding, and life will ask us more poignantly than we might imagine: "Who are you *really*, wanderer?"

REFLECTION

**What is my relationship with nature at the present time?

**When I take myself out into nature's solitude, what do I experience? (Take plenty of time and space to reflect on this question, and do not be afraid to include "negative" experiences.)

**What feelings do I experience as I consciously enter solitude in nature? Include negative emotions.

**What kinds of sensate experiences do I have in natural solitude? Sights? Sounds? Smells? Textures?

**What do I experience when I leave solitude in nature and return to civilization?

**What do I feel when I think about learning survival skills? How do I feel when I am learning and practicing them? What additional skills might I want to learn? Why? What skills do not interest me? Why?

**How are you different in the midst of natural solitude from how you think, feel, and behave in the midst of civilization?

A powerful exercise offered by Alberto Villoldo in his book *The Four Insights: Wisdom, Power, and Grace of the Earthkeepers*[32] called "The Query" is extremely useful for experiencing who we are at a very deep level.

The Query

Sit comfortably in your favorite chair and dim the lights in the room. Light a candle if you wish, but make sure that you're in an absolutely quiet place because you want to listen to the chatter of your mind. Close your eyes and begin to take deep, regular breaths...count your breaths from one to ten, and then start at one again.

After a few minutes, you may notice that you're counting up to 27 or 35, as the mind becomes absorbed with what you need to do later in the evening, what you failed to do at work, or how upset you are with someone. Or perhaps there's a tune playing inside your head.

Bring yourself back to counting your breaths. Now ask yourself, "Who is angry?" "Who is late?" "Who is breathing?" and then, "Who is it that is asking the question?" Be still, and observe what happens when you ask this.

You can also ask this question throughout the day when you are not sitting in meditation.

For sixty years
I have been forgetful
every moment, but
not for a second
has this flowing
toward me
stopped or slowed.
I deserve nothing.
Today I recognize
that I am the guest
the mystics talk about.
I play this living music
for my Host.
Everything today
is for the Host.

~Rumi~

NOTES

NOTES

Chapter 7–Savoring the Moment, All Things Are Slipping Away

One thing all human beings have in common is loss.
~Malidoma Somé~

As we know, most humans, especially Americans, resist dealing with death. However, even in "ordinary" times, the willingness to think and feel about one's own death can be useful in preparing us for it. The end of life as we have known it makes the contemplation of death on a number of levels not only useful but necessary, for after all, collapse itself is nothing if not a monumental death event— the death of civilization and the roles we play in it as well as what we receive from it, however evil, benign, or glorious that may be.

Storyteller and mythologist, Michael Meade, encourages us to begin the encounter with our final death by consciously opening to all of the "little deaths" in our lives—or if you prefer, as stated above—the lesser collapses. Each one offers a lesson that will prepare us for our final exit from our embodied earth existence. Stephen and Ondrea Levine have written and spoken profusely about the experience of death, and in a 1998 interview with Randy Peyser, Stephen Levine states[33]:

The basis of conscious living and conscious dying are precisely the same thing. They have

93

> *to do with paying attention to the moment as it unfolds, and trying to meet it with as much mercy and compassion as possible. If we're being conscious in this moment then we'll be conscious in that moment. The less we're being conscious now, the less we're being heartful now, and the more difficult our death bed might be.*

Levine is asking us to embrace a daunting task—keeping our hearts open as we face death in its many forms. As we sit in the cauldron of loss in 2009—as we witness climate change eradicating much of earth's geography, and as scientists tell us that we may have passed the point of no return and that there is virtually nothing we can do to prevent climate chaos—as we witness the extinction of some two hundred species per day, as we watch the global economy and basic societal institutions crumbling, as we observe gargantuan chunks of open space that have been devoured by "development", as we notice how "growth" has eviscerated our external world, and as "growths" claim the lives of our friends and loved ones with myriad forms of cancer and other debilitating diseases, as unemployment increasingly becomes the rule, rather than the exception to it, and as millions become homeless and unable to acquire healthcare, we feel ourselves surrounded by and immersed in a plethora of deaths.

Collapse unmistakably signals that much of what we have held dear in civilization is in the process of irreversible demise. The planet, the simpler lifestyles with which we grew up, the non-human and human worlds, the climate, the ecosystem—all are in a process of dying. Resources are shrinking dramatically, and humankind's irreparable soiling of its own nest is exacerbating at vertiginous speed.

Being surrounded by relentless manifestations of death, we are forced to contemplate our own death. If indeed, as numerous spiritual teachers have reminded us, we are inextricably connected with all of life, then death in one corner of the universe affects our immediate world on some level. If our bodies are part of earth's body, then our own bodies cannot escape the injury of poisoned land and water, the extinction of species, the ravaging of rain forests, and the toxicity of our air and food. It all feels so hopeless, and indeed, on one level, it is.

So now we enter new territory because the moment I suggest confronting one's hopelessness, I am also inviting us into deeper layers of the psyche which is the Greek word for *soul*. At that point we are under the radar of theories, facts, and even paradigms. We are brushing against our deepest terror, our most excruciating grief, and our billowing, frothing, fulminating rage. Suddenly, we are confronting our human limits, and in fact, our very own death. Yet until we can affirm that the planet is in a death struggle both literally and metaphorically, and until we can adopt the attitude that we are doing nothing less than inhabiting our days and hours in a funeral procession, we will kick and scream for hopeful solutions.

This is conscious preparation for death, and all those who are willing to embrace the reality of collapse are hospice workers for ourselves and the world. There isn't much time left, and every moment is a gift to be savored, smelled, tasted, touched, and caressed.

Someone has said that death is a place in the middle between birth and rebirth. In terms of literal death in this lifetime, we only experience it once, and whether it is our own death or the death of planet earth, it is as sacred as the moment of our birth. It is everyone's right and privilege to defend against death and in so doing, opt for disempowerment. But as we have seen in previous chapters, our power is not to be found in denial, but in conscious exploration. For this reason the attitude of the hospice worker is the one that will best serve us as we walk willingly forward in the funeral procession of our own and the planet's death.

The earlier subtitle of this book, "restoring life on a dying planet" was not chosen because I believe that collapse can be prevented or that the planet and its expiring inhabitants can be resuscitated, but rather because I hold that by opening to the death that is all around us, we make rebirth possible. I do not know what form that transformation will take.

What I do know is that paradoxically, openness to death enables opportunity for new life.

While it is appropriate and necessary to mourn our incalculable losses, it is equally important to be grateful for what we do have. For example, every delicious meal consumed should be done with an attitude of gratitude that we still have access to a supply of wholesome and abundant food; every drink of clean, clear water should be savored with gratitude that it is available to us; every time we embark on a short trip by car or on public transportation, we have the opportunity to give thanks that the energy for doing so is still available; every time we lie down to sleep in our beds, we have the opportunity to give thanks for shelter and a safe place to rest; every time we put on clothes and shoes we have the privilege of giving thanks for the people who made them and the convenience of not having to make our own, even as we contemplate learning the skills to do so.

Perhaps more than any other asset in our lives we most take our health for granted, but as we march in the funeral procession of collapse, it behooves us to give thanks for our five physical senses, our ability to walk and move about in our world. For the most part we still have access to medications and over the counter remedies, and we have the luxury of educating ourselves about natural healing techniques before full-blown collapse makes such learning extremely difficult if not impossible.

The list goes on, and while I could continue elaborating, you are probably getting the idea. Nothing should be taken for granted. In fact, the Benedictine, Buddhist-oriented Brother David Stendl-Rast in his wonderful little book *Gratefulness: The Heart Of Prayer,* plays with the expression "taking for granted". We constantly engage in the unconscious activity of taking things and people for granted, but we have the opportunity to consciously "take someone or something for granted", by affirming that they have been *granted* to us, and that in fact, all people and things in our lives—all of our abilities and our very lives are gifts. When we do not perceive them as *granted*, we are asleep and unconscious.

The connection between gratitude and meaning is succinctly summed up by Stendl-Rast: "Gratefulness is the inner gesture of giving meaning to our life by receiving life as a gift."

If we want to increase our gratitude sensitivity, then we need to become "heartful" as Stephen Levine implies in his words at the beginning of this chapter. Stendl-Rast adds, "The organ for meaning is the heart." Gratitude can strengthen the heart because gratitude is fundamentally about relationship, that is, a connection with everything we are grateful for, which in my experience inevitably leads to a connection with the sacred as the ultimate source of life.

I can think of no poem that more thoroughly captures the spirit of gratitude of which I am speaking than Mary Oliver's "Messenger":

My work is loving the world.
Here the sunflowers, there the hummingbird —
equal seekers of sweetness.
Here the quickening yeast; there the blue plums.
Here the clam deep in the speckled sand.

Are my boots old? Is my coat torn?
Am I no longer young, and still not half-perfect? Let me
keep my mind on what matters,
which is my work,

Which is mostly standing still and learning to be
astonished.
The phoebe, the delphinium.
The sheep in the pasture, and the pasture.
Which is mostly rejoicing, since all ingredients are here,

Which is gratitude, to be given a mind and a heart
and these body-clothes,
a mouth with which to give shouts of joy
to the moth and the wren, to the sleepy dug-up clam,
telling them all, over and over, how it is
that we live forever.

While some of the above notions about gratitude may stir warm feelings within us, they are profoundly difficult to practice. When we contemplate collapse, we are staring into the end of the world as we have known it, and all of the adversities mentioned in earlier chapters of this book glare back at us formidably. If anything, they make us want to stiffen the upper lip, suit up, armor ourselves, and rationally problem-solve our way through the frightening challenges ahead. Yet much "heart work" must be done. In Chapter One I noted Victor Frankl's comments on finding meaning in the midst of a Nazi death camp—a hellish place to imagine being heartful. Yet a certain degree of open-hearted sensitivity was required in the face of ghoulish brutality, and those who had completely anesthetized their feelings were less likely to survive.

One of the most powerful ways of becoming heartful is to contemplate one's own death. The Dali Lama states that he spends about one hour daily thinking about his own death. Those of us in the second half of life are generally more willing to do this, but familiarity with physical death is not the only death to which I'm referring. As noted in Chapter 5, collapse strips us of who we think we are—a kind of death that may in some cases be more painful than physical death. That death is the demise of the human ego which collapse or any other initiation invariably evokes. On some level, all of us will die in the process of collapse; some will lose their physical lives, but everyone will have his/her ego shattered.

Commenting on the death of the ego, Stendl-Rast says that "We might be killed in the process, but we come out of this experience more alive....There is no telling how many times in the course of a lifetime we may have to go through this process of *creative dying* [emphasis added]. The more creatively we live, the more often we shall have to die, I suppose…There is no growing without dying to what we have outgrown…If we have the courage to let go of each other, this death experience will become creative." And I might add, if we have the courage to let go of our ego definition of who we are, an ego-death can be unimaginably creative. Collapse will force nations, communities, families, and individuals to experience ego-death. That reality is not optional; what *is* optional is the degree to which we consciously participate in the purpose that the initiation seeks to accomplish.

During the spring and summer of 2008 as I prepared for and departed on my journey of relocation to Vermont, I became riveted by the writings of Eckhart Tolle's *The Power of Now* and *A New Earth*. Driving 3,000 miles from the New Mexico desert to the lush, rolling hills of Vermont, I was reassured byTolle's podcasts that among the myriad motivations for making this particular move was my openness to ego death and a desire for the transformation of consciousness. I was experiencing yet another chapter in a lifelong journey of waking up—not only to the events of the world around me and their gravity, but more importantly, to my essential nature and the fact that it is not separate from anyone or anything in the universe.

While any individual can experience this anywhere—while I could have continued the awakening process in the American Southwest as well as the Northeast, circumstances were calling me to a different part of the country, to a different climate, and to a different subculture. As is often the case when we relocate to some other part of the planet, I felt disoriented as if not knowing exactly who I was. Of course, this is exactly what I needed to feel because my ego identity was shifting, and layers were being peeled off. In the process I grew more intimately connected with my essence and the awareness that I was being called to a place more conducive to the ego-shedding, essence-embracing adventure. I was re-discovering on a new level, as Tolle says, that "you don't live your life, but life lives you. Life is the dancer, and you are the dance."

My ongoing work as a human being who longs to be awake, is to facilitate the death of my ego and the birth of the greater self within and around me and to constantly experience my relatedness to all that is. This is the purpose to which, in one way or another, we are all called, and it may be that collapse is necessary in order to reveal to us unequivocally, that reality.

I have come to believe that whether he is aware of it or not, Eckhart Tolle may be preparing all of humanity for collapse and the transformation it will evoke. In *The New Earth* he writes: "A significant portion of the earth's population will soon recognize, if they haven't already done so, that humanity is now faced with a stark choice: Evolve or die." On one level, Tolle is speaking of physical death, but on another level, he knows well that the evolution of consciousness does not preclude physical death. As collapse exacerbates, many individuals may well evolve *and* die; however, the more one identifies with the ego, the more daunting will be the experience of death. The less ego-identified one is, the more profound may be one's awareness of evolution's magnificent design which Clarissa Pinkola-Estes names the *life/death/life* process.

As hospice workers for those around us, we can assist them in consciously releasing their egos and the trappings of civilization to which they have become attached. And, as with hospice work in the case of literal physical death, assisting another in her/his ego-death can facilitate our own process of surrender. We know not what we will be called upon to feel or do, nor do we know what emotions and circumstances other individuals in our lives will encounter. What we must know is that opportunities for learning and teaching surrender will be omnipresent.

REFLECTION

**Who are the specific people in your life for whom you feel most grateful? Why?

**What gifts or "blessings" in your life are you aware of feeling grateful for?

**What people and blessings would be most difficult to part with? Why?

**From what you have read thus far and from your own inner work, what parts of you might have to die in order for your personal transformation to occur?

**How is your heart? Where do you feel your heart closed or closing? In what circumstances and around which people? Why? When do you notice your heart opening? What causes it to close again?

Suggested Practice:

**Spend an entire day allowing yourself to feel grateful for everyone and everything you encounter, including the gifts mentioned in this chapter. You are likely to forget quite easily. When you do, just remind yourself to resume the practice. At the end of the day, journal about your experience.

**Spend another day noticing the condition of your heart—when it closes, when it opens, when it feels vulnerable and tender, when you aren't feeling it at all. At the end of the day, journal about your experience.

NOTES

Notes

Chapter 8–Beauty, the Gift That Visits and Vanishes

Let the beauty we love be what we do.
There are a hundred ways to kneel and kiss the
ground.

~Rumi~

The late Irish poet and former priest, John O'Donahue wrote a lovely book entitled *Beauty: The Invisible Embrace*[34]. Until reading it I hadn't realized that so much could be stated about the topic. Unlike other species, it seems that humans have a unique capacity for appreciating beauty, but frequently upon discovering its loveliness, we attempt, as we do with so many other things, to control it—to make it happen when we want it to. Yet, O'Donahue reminds us that "When we walk on the earth with reverence, beauty will decide to trust us." So it would seem that as with nature, we must be willing to cultivate a relationship with beauty, and to do so we must be willing to allow it its own timing and fickle consent to reveal itself. In a world permeated with greed where power and control are ubiquitous, it may be very difficult to allow beauty to come to us rather than attempting to go to and capture it.

"Beauty loves freedom," says O'Donahue, so humility in its presence is a pre-requisite. Words like "humility" and "reverence" in relation to beauty imply that there is a sacredness in its origin and in its "revelation" to us, that abides at the heart of life.

Sadly, one of the most likely casualties of collapse is beauty. Much of its grandeur is rapidly slipping away. Within the next few decades we are likely to witness fewer unadulterated blue skies, clear starlit nights, lush green forests, rolling verdant meadows, blue-green rivers and lakes,--fewer flowers, trees, snow-capped peaks, or handsomely-marked birds or animals.

Many avenues of beauty on which we have relied in the past to grace our five physical senses with extraordinary sights, sounds, smells, tastes, and textures may be less accessible or completely inaccessible. Consequently, it will be necessary to acquire beauty more consciously in less likely places and with more intention.

Yet in the ugliest of circumstances beauty may be found. Even concentration camp inmates during the holocaust were able to find beauty. Some like Fania Fenelon and the Musicians of Auschwitz had access to musical instruments as depicted in the 1980 television movie "Playing For Time." Others had to secretly create makeshift instruments, or drawings, or sing alone or in groups. Yet by whatever means they created beauty, it added to the meaning of which Frankl wrote, and it nurtured the mind and emotions when no other solace could be found.

Collapse is likely to leave the landscape bereft of beauty and punctuated with large pockets of visual, as well as other forms, of pollution. Even as I write these words, suburbs are becoming slums or abandoned ghost towns as home foreclosures leave behind unprecedented blight; what few farmlands have not been developed may lie fallow for long periods and deteriorate into vast acreages of weeds; forests may be increasingly ravaged with fires as global warming exacerbates; rivers may become disgusting lagoons of sludge and raw sewage. A December, 2008 story in the *New York Times* "Back At Junk Value, Recyclables Are Piling Up" reported that the economic downturn has decimated the market for recycled materials which are accumulating on streets, on the landscape, and in landfills.

We cannot know the extent to which energy depletion will prevent us from hearing beautiful music on radios, CD players, or computers. Musical instruments should be included in our cache of preparatory items, especially drums which are so intimately connected with nature.

Above all, we must treasure our own voices and those of others. "All holiness," Donahue writes, "is about learning to hear the voice of your own soul." Poems and stories that we have committed to memory will be invaluable conduits of beauty and truth which O'Donahue insists "are internally linked" and "...if we choose to journey on the path of truth, it then becomes a sacred duty to walk hand in hand with beauty." A sacred *duty*? Yes, how not, in the throes of the demise of industrial civilization? A powerful story or poem well-spoken can heal, unite, and inspire like no other form of communication. These can also evoke grief, rage, ire, and other so-called "negative" emotions. But lest we become too somber about voices and music, let us remember that they are also invaluable for celebration and stress release. Combined with dance, voice and music can solidify the community in sacred ritual, merriment, and conviviality as they have for thousands of years among our tribal ancestors.

Perhaps no other quality welcomes or accesses beauty as profoundly as the human imagination. It is a direct conduit to beauty and the divine within. Out of our imagination we create, and when we create, we are participating in the sacred. O'Donahue lists the gifts that imagination brings:

- Imagination is the eye for the inner world.

- Imagination knows well the shadows and troughs of the world, but it believes that there is more, that there are secret worlds hidden within the simplest, clearest things.

- It retains a passion for freedom.

- It keeps the heart young.

- It awakens the wildness of the heart.

- It causes us to become interested in what might be rather than what has always been.

- It offers wholeness: heart and head; feeling and thought come into balance.

- Imagination offers revelation, never blasting us with information but coaxing us into a new situation with gradualness.

- Imagination works through suggestion which respects the mystery and richness of a thing.

- It has a deep sense of irony.

- Imagination creates a pathway of reverence for the visitations of beauty.

The final chapter of O'Donahue's *Beauty* is entitled "God Is Beauty." While some may be put off by the word "God", this chapter is so exquisite, so stunningly "beautiful" that it bears reading and re-reading many times. I have underlined almost every sentence in it, but at this juncture, one leaps out: "Spirituality has to do with the transfiguration of distance, to come near to ourselves, to beauty, and to God." In other words, beauty is that magnificent and mysterious bridge to nature, to ourselves, and to the sacred. If our work is to further open our hearts, allow our egos to die, and fully surrender to the greater self, then beauty is the consummate facilitator of that process.

In preparation for collapse, we can savor the myriad forms of beauty in our lives and cherish them with utmost reverence and honor, aware that they will not always be with us. As the seventies singer and composer Carly Simon wrote, "These *are* the good ole' days." In five years, a delicious meal like the one that a friend may have just cooked and that we consumed with gusto may not be available, and what is more, our friend may not be present with us. The cozy, warm, soft bed in which we sleep tonight may not be the place where we are forced to sleep in a post-collapse world. Many people, creatures, necessities, and amenities that we now enjoy may not grace our lives and our homes in years to come. Now is the time to revel in their presence in our lives. Now is the time to delight in every moment of beauty, joy, love, conviviality, playfulness, intimacy, sensuality, and fullness of life with which we are graced in these present moments.

Mary Oliver in "Blackwater Woods" exquisitely describes our spiritual and life practice as we inhabit our days in this and every year:

Look, the trees
Are turning
Their own bodies into pillars
Of light, are giving off the rich
Fragrance of cinnamon
And fulfillment,

The long tapers of cattails
Are bursting and floating away over
The blue shoulders

Of the ponds,
And every pond,
No matter what its
Name is, is nameless now.

Every year
Everything
I have ever learned in my lifetime

Leads back to this: the fires and
The black river of loss
Whose other side is salvation,
Whose meaning none of us will ever know,

To live in this world
You must be able
To do three things:
To love what is mortal;
To hold it against your bones knowing

Your own life depends on it;
And, when the time comes let it go,
To let it go.

REFLECTION

**What forms of beauty are especially dear to me—so much so that I can barely stand to think of parting with them?

**Where do I find beauty in my daily life? In my community? In my home? In my work?

**How does beauty affect the condition of my heart? Do I need to indulge in more beauty in order to nurture my heart? If so, where and how can I do so?

**How does beauty enrich or complement my sense of the sacred?

**How can I make my home, my workplace, my community more beautiful?

**What is the connection for me between beauty and gratitude?

**As collapse unfolds, we are likely to increasingly lose more of the natural beauty around us. Take time to journal about this, particularly writing about your feelings when you think of this possibility. Perhaps the writing that comes forth will be attended by deep feelings. If so, please take time and space to allow and honor them. You may find it helpful to journal about this topic more than once. Or, given the topic, it may be equally helpful to express emotions by drawing or creating visual images.

NOTES

NOTES

CHAPTER 9–SCARCITY, ABUNDANCE, AND ALTERNATIVE FORMS OF EXCHANGE

To laugh often and much; to win the respect of intelligent people and the affection of children; to earn the appreciation of honest critics and endure the betrayal of false friends; to appreciate beauty; to find the best in others; to leave the world a bit better whether by a healthy child, a garden patch, or a redeemed social condition; to know even one life has breathed easier because you have lived. This is to have succeeded.

~Ralph Waldo Emerson~

Surrounded by or reveling in beauty, we rarely feel the impact of scarcity. In those moments our lives feel abundant, sometimes as if we need nothing more than the beauty by which we feel magnetized. Countless artists and musicians throughout human history who were pre-occupied with creating beauty also lived in abject poverty and were frequently unknown and unappreciated for their extraordinary gifts during their lifetimes. Perhaps beauty was their "fullness" and "sustenance". Obviously, they couldn't eat beauty, nor did it pay the bills, but they seemed to have had a remarkable ability to endure scarcity in the midst of the stunning creativity they were exuding.

A collapsing and post-collapse world is one of varying degrees of scarcity. In the current milieu some individuals believe that trading in their SUV for a hybrid vehicle is a sacrifice; others refuse to rent instead of buy because they feel they must have the American dream of owning a home despite the financial angst that home ownership in today's economy often entails. Most of these individuals cannot imagine a world in which they might be homeless, forced to walk or ride a bicycle and live without a petroleum-powered vehicle—or, not know where their next meal is coming from.

Unlike citizens of the developing world, Westerners—especially Americans, have been inculcated with Puritan values. That is, if one works very hard, plays by the rules, and pays his bills, he will prosper. According to the Puritan ethic, poverty is a glaring indication that one is not living a "good Christian life." While most individuals in the United States today would minimize the Puritan influence in their value system, that ethic is alive and well in deeper layers of the American psyche. Even amid today's bankruptcy and foreclosure epidemics in America, people in the early or middle stages of declaring bankruptcy are usually extremely uncomfortable talking about it. Understandably, they are not proud of their circumstances, but inordinate shame is associated with bankruptcy in this culture, and I believe that the Puritan work ethic may be near the top of list of reasons for this.

I grew up in an extremely rigid fundamentalist Christian family, but having "money issues" was a far greater evil than committing sexual sins—and for fundamentalist Christians, almost nothing is more shameful than "transgressions of the flesh." During the past three decades, the so-called New Age religions have not alleviated the Puritan ethic in relation to scarcity, but in my opinion, reinforced it. The 2006 book *The Secret* based on "the law of attraction" admonishes readers that if they are not wealthy, it is because their attitude is creating scarcity and promises that "The Secret" can teach them how to attract abundance.

For the most part, purveyors and adherents of *The Secret* are members of America's white middle class. While they may be having difficulties making their mortgage payments, paying off credit card debt, or taking vacations, they have almost no comprehension of the grinding poverty and suffering rampant in developing countries around the world. It is precisely that kind of scarcity that is becoming epidemic in a collapsing America, and individuals who do not take collapse seriously, who insist on denying or minimizing it, will be tragically at risk when they find themselves in the throes of it.

Dmitry Orlov in *Re-Inventing Collapse: The Soviet Example And American Prospects* states that "To keep evil at bay, Americans require money," and he devotes a significant portion of his book to suggesting ways that people in a collapsing society can circumvent the money system and develop other resources for getting their fundamental physical needs met. Specifically, Orlov states:

If life without money is to become normal for most people in the US, then it seems inevitable that the flow of humanity will become bifurcated. Those who are most helpless will find themselves on the inside, in institutional settings such as jails, asylums and hastily organized camps for the internally displaced, kept alive while the institutions hold together and the supplies last. Those who are resourceful will find ways to remain on the outside, and may find themselves pursued and persecuted as terrorists while the institutions hold together and supplies last, but eventually they will be left alone as the supplies needed to continue persecuting them run low. Some clever people are sure to find ways to work as conduits between the two worlds, living undercover among the outsiders to obtain intelligence for their institutional masters, but really just looking out for themselves and their friends, and conveying people and supplies back and forth.[35]

Individuals who have been taking collapse seriously enough to make conscious preparation for it will not be battered and disoriented by the scarcity of collapse because they will have been psychologically and financially preparing for it. That does not mean that they will experience no hardships. Of course they will, but most of those individuals will have, over time, tempered their needs in accordance with the fine-tuning of their values resulting from educating themselves about collapse and preparing themselves for it emotionally and spiritually. Moreover, many are and have been organizing local economies around food security, the exchanges of goods and services, community currencies, and a much more cooperative style of living.

As we know, the developing world has always relied on resource-sharing and community efforts to make survival possible. In most of those cultures, the extended family and community are primary, unlike the American lifestyle based on individualism and the nuclear family. Collapse will indeed necessitate reliance on friends, neighbors, and members of the community in ways that Americans inculcated with individualism cannot possibly imagine. Only two options exist: One will either adapt one's individualism to cooperative living or one will not survive.

For this reason, two preparations that anyone who takes collapse seriously must make are: 1) Learning survival skills, 2) Participating now in relocalization efforts in one's community or relocating to an area where this is feasible.

In previous chapters I have mentioned a number of skills that would be useful to learn in preparation for collapse. Individuals as well as families can and should learn them, and although each person will have preferences about which skills they want to learn, it would be useful for everyone in a family to learn at least one skill. With the title and intention of this book in mind, I do not wish to offer more lists of skills to be learned, but rather frame the notion of learning them in the context of spiritual practice and soul-enhancement.

Although learning canning or woodworking is not the same as sitting in the grass beside a river with one's back against a tree on a luscious, golden autumn day contemplating one's connection with nature, learning survival skills has the potential for deepening our relationship with the natural world. One of the tragedies of modernity is the lack of awareness, particularly among the youth, regarding the origin of so many foods, services, and products which we take for granted on a daily basis. One's world shifts slightly but significantly when having the experience of milking a cow as opposed to pulling a carton of milk off a grocery shelf and tossing it in a shopping cart. The satisfaction one feels in the body and mind when completing a chair or bookshelf for which one has cut the wood, carved it into useable sections, finished the parts and then assembled them into a piece of furniture is profound.

Learning survival skills is intimately connected with expressing our creativity. We learn to make a product, plant and harvest a garden, can fruits and vegetables, harvest and brew medicines from herbs, sew a shirt or pair of pants from scratch, or hunt and butcher a game animal—all of these skills creating something that did not previously exist and which meet fundamental needs that will not and cannot be met by anyone but ourselves or our community in a post-petroleum, crumbling world. But the satisfaction we enjoy from these experiences is not merely for our benefit alone.

Practicing relocalization now in our communities could be a labor of love through which we can become delightfully familiar with, before a collapsed world forces us to adapt to, this cooperative arrangement. The spirit of relocalization is one of sharing and working together with our local community to insure that everyone prospers. The Business Alliance For Local Living Economies (BALLE)[36] is a treasure-trove of resources for creating and strengthening local economies. BALLE's website lists its major networks around the nation, but in addition, relocalization is occurring in hundreds of other locations as awakened individuals realize that the future of their communities depends on it. More recently, the Transition Town movement, which originated in the UK, has become a worldwide movement and offers a no-nonsense blueprint for transitioning our communities to a post-petroleum, post-monetary lifestyle. Specific strategies are outlined in the *Transition Handbook* by Rob Hopkins and will be noted again in Chapter 20 of this book.

At this writing in 2009, a perfect storm of economic depression is building in the United States. Millions across the nation are losing their homes, jobs, pensions and retirement savings, and navigating foreclosures and bankruptcies. The economic crisis can only be exacerbated by Peak Oil, climate change, loss of healthcare, severe inflation or deflation, a mindboggling U.S. deficit, stolen pension funds, millions of baby boomers heading for Social Security and Medicare rolls, and perhaps the most formidable reality of all: the abject insolvency of the United States government.

For years, a host of economists and analysts have warned us that ultimately, the U.S. dollar will also collapse and become worthless. Some insist that currently it is only worth a few cents in terms of its real value. Few of us can comprehend how momentous "the end of money" will be, but one thing is certain: It will force us, whether we like it or not, to: 1) transact business in our local communities and, 2) do so using different standards and mediums of exchange.

This will not be unprecedented for the United States since during the Great Depression, there were thousands of local currencies across the nation. Alternative currencies, as well as barter, are likely to be used in a collapsing or post-collapse world, and this could have a number of effects. There is likely to be much confusion about what form of currency to use and how to calculate its value, and unless people are committed to cooperation instead of competition, endless conflicts may result.

Barter may be very difficult to adjust to since it will force people to appraise the value of everything they trade. And once again, as we think about "value" we have entered the territory of the soul. No longer is the issue simply economic. Enormous soul-searching will occur as people emotionally "weigh" the value of items or services traded for other items or services. Nothing could be taken for granted; everything would have new and unprecedented "worth." Obviously, those who have begun consciously preparing for collapse are even now entering it with transformed value systems that will incisively inform their transactions in a moneyless world. And once again, as we think about "worth" and "value", we are invited back to the subject of gratitude and relationship—thankfulness for all that we possess and an exploration of our relationship with it. Our spiritual practice continues to be our heartfelt appraisal of the value of all that blesses our lives, both the tangible and intangible.

REFLECTION

**What are my most prized intangible possessions? Spend some time journaling about this.

**What are my most prized tangible possessions? Spend some time journaling about this also.

**What efforts are being made in my local community to create and strengthen a local economy? What is my understanding of my local economy and how can I support it with my purchases and with my time and energy?

**If you do not live in a community that values relocalization, what would you like to do about that?

**If you have learned any survival skills, spend some time journaling about how you have felt during and after learning these skills. How has the acquiring of these skills shifted your outlook, values, and your connection with nature?

NOTES

NOTES

CHAPTER 10–LOSS--GATEWAY TO THE GOLD

*Sometimes I go about pitying myself, and all the time,
I am being carried on great winds across the sky.*

~Chippewa Song~

The end of the world as we have known it is replete with goodbyes. On the one hand we tout the adoption of new lifestyles, but that does not erase the memory of our former ones. As Richard Heinberg writes at the end of *The Party's Over*, his remarkable analysis of the Peak Oil crisis, "It *has* been a fabulous party." Undeniably, a world awash in oil provided lifestyles we would have never had without it—all those wonderful household gadgets, luxury cars, jetting about the world at will, and of course, endless technology: satellite TV, cell phones, computers, Ipods, plasma screens, Black Berry's, and so much more. But as Heinberg quickly adds, "From those to whom much has been given, much should be expected."

In James Howard Kunstler's novel *World Made By Hand*, the author takes us into a post-petroleum, post-collapse world and paints a poignant picture of the grief that survivors feel regarding the untold losses resulting from collapse and the gray pallor of sorrow that lies just behind the contentment that they have created for themselves by living simple, meaningful lives off the electrical grid and without the former trappings of civilization. While reading the book I was profoundly in awe of Kunstler's depiction of loss in the new made-by-hand world and the emotional toll it took on its inhabitants.

A collapsing world will impose loss on everyone. Depending on our level of preparedness which is inextricably connected with the priorities we establish, there will be many goodbyes, and not all of them voluntary. We will choose to leave or be forced to walk away from certain individuals, from houses, possessions, familiar places, established routines, and many hopes and dreams. Products of civilization that we are, these losses are likely to feel excruciating or at the very least, evoke deep resentment and sorrow.

On the other hand, indigenous traditions devote a great deal of time and energy to preparing their people for life's losses because those traditions teach that loss is an enormous and necessary part of the human experience. Unlike civilized Americans, traditional peoples understand that the majority of life's experiences will be challenging or unpleasant and that their songs, stories, ritual, and established communities are gifts they can utilize to support themselves during periods of adversity and to celebrate their perseverance afterward.

In hunter-gatherer societies, transitions were incessant as the tribe constantly moved from place to place in search of food. They were extremely limited in what they could carry with them because continuous mobility was necessary for survival. Only in sedentary societies are humans able to develop the kinds of attachments to possessions and places that today often encumber us as we navigate transition.

Separations that we choose are difficult, but more so the ones that may be imposed upon us. For example, it is one thing to choose to say goodbye to a particular location where we have developed strong attachments, and quite another to find ourselves hard-pressed to obtain a glass of clean, safe drinking water in that location. Many of us are keenly aware of the disappearance of the honey bees, yet it is not so dramatic that we notice it daily. But how might it feel to witness miles and miles of burned or pillaged landscape, ravaged forests, drought-battered river beds containing only dry sand, or hundreds of acres of housing tracts that have become ghost towns or furtive hide-outs for the homeless?

On a more personal level, what will it be like in a collapsing world, or for that matter, any world, to notice our own beauty, health, and vitality slipping away? In later chapters we will consider specific challenges such as aging, illness, and death, but for now, let's ponder the issue of loss itself with emphasis on appreciating its purpose in our lives.

The first thing we must understand about loss is that we cannot evolve into mature adults without it. All children must lose their innocence and be wounded in the process—another reason for a ritual of initiation during the adolescent years. Loss defines us, tempers us, tests us, transforms us, and it never ceases to do so throughout our lives. In fact, many spiritual teachers, including psychologist Carl Jung, have suggested that as we become more conscious and wizened, our losses do not become smaller, but rather larger. Certainly, this flies in the face of the Puritan ethic and its New Age manifestations which imply that as we become "more evolved", life's challenges grow smaller or even cease to exist. The Jungian perspective holds that as more of our essence comes to consciousness, we therefore have more internal resources with which to meet the challenges, and as a result, the challenges are likely to increase is size or frequency or both.

Naturally, encountering adversity at sixty is going to be very different from the adversity experienced at sixteen, but from the perspective of soul development, conscious cultivation of the soul equips the individual to more skillfully navigate the complexity of adversity in later years. While the aging process is typically challenging for most people, its intensity is likely to be magnified by the collapse of civilization.

Secondly, our pivotal task in consciously experiencing loss is to grieve it, and I cannot over-emphasize this. We all know individuals who have never grieved monumental losses in their lives, and we witness the results of that—illness, anger, fear, depression, isolation—the list of consequences is virtually endless.

As Sally Erickson states in "What A Way To Go", the only sane response to the devastating losses with which our planet and our lives seem replete is deep grief. Unless we allow ourselves to consciously grieve these losses, we risk acting out the grief in some other way that will be harmful to ourselves or those around us.

When I lived in the San Francisco Bay Area in the mid-nineties many of us who attended workshops with Malidoma Somé learned grief rituals practiced by his West African Dagara tribe. Since that time a group of my Bay Area friends have met annually in the season of the Day of the Dead to participate in a grief ritual. The day is devoted to building an altar to which during group drumming and singing, the participants bring their grief and express it at the altar through weeping, wailing, or praying. These kinds of rituals are extremely powerful, not only for the individual, but for the entire community, and I believe they can provide an invaluable and necessary opportunity to grieve the losses that collapse will certainly manifest and multiply.

Nevertheless, we need not wait for a group grieving ritual to feel our losses. We can take ourselves to cherished and sacred places in nature or in our homes and allow the tears to flow. We can journal, write poetry, and express our grief through art and music—but we must grieve because some part of us already knows that when two hundred species per day have gone extinct, some aspect of ourselves has also gone extinct. When forests are raped by clear-cutting, so are we. When rivers are turned into septic tanks, some part of us has been reduced to sewage. If we believe anything at all of what the great ecologists like Daniel Quinn have told us, we are inextricably, inexorably connected with what we call "nature" and "resources." Nature's rivers flow in our veins; the elements of its soil reside in our bodies; from moment to moment we bring its air into our bodies and exhale that air back into nature. We *are* nature, and nature *is* us.

Roll Call
By William Stafford

Red Wolf came, and Passenger Pigeon,
The Dodo Bird, all the gone or endangered
Came and crowded around in a circle,
The Bison, the Irish Elk, waited
Silent, the Great White Bear, fluid and strong
Sliding from the sea, streaming and creeping
In the gathering darkness, nose down,
Bowing to earth its tapered head,
Where the Black-footed Ferret, paws folded,
Stood in the center surveying the multitude
And spoke for us all: "Dearly beloved," it said.

REFLECTION

**What losses in my life stand out in my mind and my heart as the most difficult? Spend some time journaling about each one.

**Have I consciously grieved any of these losses? If so, how have I done this? If not, what has prevented me from grieving them? Do I need to grieve them more deeply?

**When I grieve, what do I experience? Where and how do I prefer to express my grief?

**Are there people in my life with whom I can share my grief? Who are they? As I share my grief, what do I need from them?

**To what extent have I allowed myself to grieve for the losses of the earth? What are some of the planetary losses that particularly stir my grief? You may want to journal or draw thoughts, feelings, or images as you ponder these questions.

NOTES

NOTES

CHAPTER 11–THE DAUNTING REALITY OF POLITICAL AND SOCIAL REPRESSION

In Germany they came first for the Communists, and I didn't speak up because I wasn't a Communist. Then they came for the Jews, and I didn't speak up because I wasn't a Jew. Then they came for the trade unionists, and I didn't speak up because I wasn't a trade unionist. Then they came for the Catholics, and I didn't speak up because I was a Protestant. Then they came for me, and by that time no one was left to speak up.

~Pastor Martin Niemoller, Arrested by the Gestapo, July 1, 1937~

Many of us have heard or read numerous times this quote from the famous German pastor. We know that when we do not perceive the struggles of other oppressed groups as our own, we eventually wake up to discover that our struggles and theirs are identical.

During 2007, two different women named Naomi released books which became best-sellers and which profoundly impacted me. The first was Naomi Wolf's little book, *The End Of America: Letters To A Young Patriot* which painstakingly explained the ways in which the United States is becoming a fascist state. Almost simultaneously, a much larger book with a more dense content, Naomi Klein's *Shock Doctrine: The Rise of Disaster Capitalism*, rolled off the press informing readers of the shock tactics used by the United States on other nations and on its own citizens for the purpose of maintaining implacable political, economic, military, mental, physical, domination. In the book Klein reveals a number of grotesque forms of physical and psychological techniques of torture and terrorism used by the American intelligence community, including the use of electroshock therapy on individuals and the calculated use of terrorist violence on communities to repress dissent and manipulate persons and groups in a desired direction.

Of all of the chapters in *Sacred Demise*, this one has been the most difficult for me to write as I find myself incessantly unwilling to sit down and face the horrors of a society that has become a police state where people fleeing cities and establishing communities in rural areas might be sought out for repression, intimidation, or worse. In reality, the "two Naomi's" haven't helped assuage my dread of writing this chapter; they've only fanned the flames of it.

Some of my friends who are better-researched in energy depletion than I insist that in the throes of collapse, Peak Oil will not permit government sweeps of rural areas in order to round up dissidents or those living alternative, cooperative lifestyles. Supposedly, the powers that be will have bigger fish to fry and much less oil, no pun intended, to do it with. The argument seems plausible, and I can only hope that its proponents are correct.

How increased repression of American society will play out is not something about which I care to speculate; I am more concerned with how I and other individuals who have been incisively reading the tea leaves of collapse for some time can avoid the horrors of imprisonment, torture, or execution.

Many collapse-watchers are making plans to expatriate to other countries, and many others have already done so. Wherever one might be thinking of relocating, it is important to understand that nowhere on earth will be entirely "safe." While some locations may experience less repression than others, climate change, energy depletion, and economic collapse will be happening and influencing events everywhere.

During the late 1980s and early 1990s as the Soviet Union collapsed, political repression did not affect the majority of Soviet citizens, but many other realities did. Some of the most thorough and insightful writings on the collapse of the Soviet Union come from Dmitry Orlov, in his series of articles, "Post-Soviet Lessons For A Post-American Century"[37]. Orlov describes the former Soviet Union during and after collapse, in terms of shortages, lawlessness, and generalized chaos on all levels of society. In the collapse of that society, the military and law enforcement virtually ceased to exist, thus eliminating the likelihood that organized details of police or military would be roaming the countryside, rounding up dissidents or groups of people living off the land who were simply attempting to survive. What was unleashed during collapse was not state repression but anarchy and a burgeoning black market economy attended by a defunct monetary system. Everything Orlov has written about the collapse of the Soviet Union is fascinating and instructive because the unraveling of the American empire is likely to proceed similarly.

Nevertheless, it would be naïve to dismiss police state repression--a formidable reality with which we may need to contend as collapse unfolds. In late 2008, a number of media entities including the *Phoenix Business Journal* published a report by the Army War College which stated that since economic crisis may lead to "civil unrest such as protests against businesses and government or runs on beleaguered banks" Pentagon resources and troops would be used to restore order. Concurrently, my website, *Speaking Truth to Power,* posted a story from the *U.K. Independent*, "Protectionist Dominoes Are Beginning to Tumble Across the World" which reported widespread protests and civil unrest in Greece, Russia, and China as a result of global economic meltdown.

During the eight years of the ultra-repressive Bush administration we have witnessed dramatic instances of out-of-control police behavior, particularly with the use of the taser. In September, 2007 the mainstream media carried incessant coverage of a University of Florida student tasered by police for asking an embarrassing question to guest speaker, Senator John Kerry, about whether or not he handed the 2004 election to George W. Bush. Shortly thereafter, mainstream media carried widespread coverage of an Ohio woman repeatedly tasered by police and then more limited coverage of the fatal tasering of a schizophrenic woman in a wheelchair as well as coverage of the tasering by police of a young autistic boy. Not surprisingly, a December, 2008 report from Amnesty International stated that between June, 2001 and August, 2008, 334 taser-related deaths had occurred.

In late September, 2007, police at the Phoenix Sky Harbor Airport restrained a hysterical forty-five year-old mother of three who was on her way to treatment for substance abuse. The woman was slammed face-down to the floor while one officer jammed a knee into her back and handcuffed her. She was then placed in a holding cell still in handcuffs and shackled. The official story of the Phoenix Police was that the woman was left alone, and when police returned, they found her dead—supposedly having choked herself while trying to escape from the handcuffs and shackling.

Passports are now required for U.S. citizens returning to the U.S. from Canada and Mexico. Passports may soon be required for all individuals flying domestically within the U.S., and the Real I.D. Act could become the law of the land in May, 2009 or shortly thereafter, requiring everyone in the U.S. to have an official I.D. card which includes all personal information about him or herself—drivers license, Social Security number, medical records, financial records, and much more. Supposedly, without the I.D., it will not be possible to obtain a drivers license, open a bank account, enter a federal building, or board an airplane in the U.S. Thus, it is becoming increasingly difficult to leave or enter the United States, and mobility within the country is also becoming more restricted. Meanwhile, President Bush reminded us that he had the power to declare martial law at any time, and the Department of Homeland Security has stated that should there be a terrorist act on the U.S., the terrorist threat level would be elevated to "red", and martial law will be inevitable.

While many people breathed easier after the election of Barack Obama to the presidency, assuming that his administration would not be as repressive as the former Bush administration, there is strong evidence that in the event of a terrorist attack in the United States or massive civil unrest, repression may not only return but exacerbate in the face of food shortages, power outages, massive unemployment, or other circumstances that could evoke collective panic and despair. It may be that post-collapse, the society will be so chaotic that police or military repression will not be feasible, but prior to that time, I believe that we can expect to see increasingly dramatic manifestations of repression and restriction on the mobility of U.S. residents and ever-more egregious violations of their privacy. As repression becomes more blatant, our fear levels will undoubtedly increase, and just living our daily lives may become exceedingly stressful.

As the paradigm of civilization becomes more threatened with extinction, its adherents will become more desperate and obsessed with keeping the old show of empire on the road. I believe that this reality is at the root of the current out-of-control behavior of law enforcement, and I expect that behavior to intensify.

As a result, many innocent people will suffer.

Considering the issues of repression, dictatorship, imprisonment, torture, and execution takes us not just into darkness but into the territory of the diabolical. All of the great slaughters of innocents in history, the Nazi holocaust, the extermination of populations, massacres, mass rape and sadistic brutality—all are heinous atrocities rooted in evil.

Immediately, the question arises: How do we survive in a milieu of repression and violence?

How do we endure? Oscar Schindler had his list; Anne Frank had her diary; Fania Fenelon had her piano and her voice. What do we possess internally that could assist us in creating similar resources? And with Victor Frankl in mind, how do we find meaning in life's horrors? Such existential issues can become overwhelming in the moment, so reflecting on them prior to a crisis may be extremely useful.

Perhaps the first place to begin is with a reminder that we live and always have lived in a world of enormous contradiction—a world where maintaining compassion in the face of evil is a seemingly impossible challenge. The essayist and short-story writer, Barry Lopez comments:

> *How is one to live a moral and compassionate existence when one is fully aware of the blood, the horror inherent in life, when one finds darkness not only in one's culture but within oneself? If there is a stage at which an individual life becomes truly adult, it must be when one grasps the irony in its unfolding and accepts responsibility for a life lived in the midst of such paradox. One must live in the middle of contradiction, because if all contradiction were eliminated at once life would collapse. There are simply no answers to some of the great pressing questions. You continue to live them out, making your life a worthy expression of leaning into the light.*

With Lopez's words we are again reminded of paradox, a notion mentioned above, which collapse will frequently force us to reconsider. It is one of the myriad philosophical and spiritual issues that a world committing suicide compels us to contemplate. When I use the word "existential" I am not referring to a school of philosophy but to our own existence and the meaning we find or do not find in that which we experience.

Victor Frankl reminds us that, "We who lived in concentration camps can remember the men who walked through the huts comforting others, giving away their last piece of bread. They may have been few in number, but they offer sufficient proof that everything can be taken from a man but one thing: the last of the human freedoms -- to choose one's attitude in any given set of circumstances, to choose one's own way."

I hasten to add that I am not concurring with those who glibly assert that our attitude determines the extent to which we suffer or that "we create our own reality." What I hear Frankl saying is that our power to find meaning in a horrific experience may be the only power we have in a particular situation, but that may be more powerful than the horror of the experience itself.

And once again, we are in the domain of the question that will not go away: Who do I want to be when_____? Fill in the blank as you choose.

In an era of political repression and social control, it will be necessary to cultivate survival skills, a strong community of friends, a wellspring of inner resources, and a great deal of stealth in the external world in order to navigate a collapsing empire's exacerbating Orwellian treachery. The writings of Dmitry Orlov provide clues regarding how we might cultivate such stealth, even as we understand that doing so does not guarantee immunity from the terrifying abuses of tyranny.

REFLECTION

**Understanding that no one enjoys thinking about evil, it may be useful at some point in our lives to have taken some time to reflect on what the word "evil" means to us. Please take some time to journal about this topic, allowing your hand to write whatever may want to come forth from it. As you write, notice what kinds of images emerge. Also pay attention to the emotions that accompany your writing.

**In what situations have you encountered evil in your life? How did you respond or react? Take some time to journal about this also.

**Who are some individuals you have known or read about whom you admire for their encounters with evil? It will probably be useful to journal about those individuals and why you admire them.

**Following is a quote from *Man's Search For Meaning*, by Victor Frankl:

> *Ultimately, man should not ask what the meaning of his life is, but rather must recognize that it is he who is asked. In a word, each man is questioned by life; and he can only answer to life by answering for his own life; to life he can only respond by being responsible.*

Take some time to sit with these words by journaling about them, by meditating on them, or by creating images in response to them. Feel free to contemplate them in a variety of ways.

NOTES

NOTES

CHAPTER 12–CREATING CIRCLES OF COMMUNITY: DEEP LISTENING AND TRUTH-TELLING

The distance between us is holy ground
To be traversed feet bare,
Arms raised in joyous dance
So that it is crossed.
And the tracks of our pilgrimage shine in the
*   darkness*
To light our coming together
In a bright and steady light.

Raphael Jesus Gonzales

Until now, much of this book has focused on the individual and his/her connection with the inner world and the outer non-human world. For some readers, that emphasis may be preferable to considering the material in this chapter--connecting with other humans in consciously creating what is loosely referred to as "community."

Hopefully, it has become obvious throughout the pages of this book that individuals in isolation cannot survive the kind of world that collapse is foisting upon us with ever-increasing rapidity and intensity. We do and will need each other. Yet at the same time, we are all wounded—as individuals and as members of the culture empire has imposed on us. Sadly, in listening to peoples' experiences of involvement in community, I hear incessant litanies of pain and disappointment with attempts to create community and resolve issues with the other members of it. When introducing the topic of community, I frequently hear sighs and see facial expressions of cynicism and despair followed by unabashed verbalizations of hopelessness around the possibility of successfully creating and maintaining community in a culture of empire.

In considering the daunting challenges of community, I believe that it behooves us to first define *community* because almost everyone has a different vision of it. By community I do not necessarily mean an intentional community or an ecovillage of individuals who share the same living space and interact with each other daily. I also do not mean a group of individuals and families who inhabit a particular town or village and see each other only once or twice a month. Perhaps the fundamental issue in community-building is not physical location or frequency of contact, but rather commitment to establishing and maintaining community. Close physical proximity is important, but that does not necessarily require living under the same roof.

Next, we must accept the reality that we cannot survive collapse in individualistic isolation. Even living in close proximity will not adequately support and protect us when some of the more dire aspects of collapse become obvious. If you have experienced disappointment with living in community, you may be thinking, "Maybe I can't live without community, but I can't live with it either." Without attempting to figure out how, it is crucial to first accept the reality that: *You cannot live in isolation, outside of community if you intend to survive.*

Once you have accepted that you need community, then the next task is to create that community around you, and that means thinking about who you love and trust and who you want to share the experiences of collapse with. This requires not only logical thought but gut-level intuition. It will be tempting to focus only on the logistics of living arrangements, how resources and tasks will be shared, preparation for crisis conditions, and other issues, but the one issue that frequently falls by the wayside is how to actually maintain the community one is creating. Since communities are almost always torn apart or dissolved as a result of emotionally-based issues, it is imperative that a commitment to working with feelings--that is deep listening and deep truth-telling, in the context of community, be given the same priority as physical survival. Why? *Because if emotional issues are not consciously addressed and worked through, they can and will sabotage the community's very existence.*

Much talk of ecovillages and intentional communities abounds among people consciously preparing for collapse, and in many the communities that have actually been created, a significant amount of time is devoted to community building--sometimes a minimum of three hours per day. One may wonder how anything else can get done when people sit in community circles that many hours. Who plants and weeds the garden? Who cooks? Who washes dishes and empties garbage into the compost?

What many living communities have discovered is that community building requires so much time that its members must have extricated themselves from the system of empire to such an extent that they have the time required to devote three or four hours per day to sitting in a circle and processing feelings and making decisions about the community's well being. What does not work well, experience tells us, is a community in which people share residence but are still chained to a system in which they must commute to exhausting jobs, return to their ecovillage, and have little or no time or energy left to do the emotional work necessary to sustain it.

What is more, every tribe, every community must develop skills for resolving conflict. Conflict will and should arise. Its absence is, in my opinion, a frightening red flag, signaling glaring dysfunction and seething cauldrons of unspoken feelings and truths that need to be told. All indigenous cultures at their highpoints skillfully navigated conflict, in fact welcomed it, as a barometer of their community's health. They also developed ever-more creative skills for addressing it compassionately and assertively. Fundamental to addressing conflict in tribal venues is a council of elders. Growing up in the tribe, one is taught from birth a profound respect for elders and the importance of deferring to their wisdom.

This is not to say that elders are infallible or somehow more than human. Indeed they make mistakes and are rarely known for their charm or congeniality. What enhances their stature is the wisdom they have gained and demonstrated to the community, which does not necessarily supersede the wisdom of other members of the tribe, but can often serve as both an anchor and a compass in navigating conflict.

I have been privileged to have opportunities in retreat settings to sit in dialog circles. At first I felt absolutely overwhelmed with the amount of emotional work that needs to be done in order for community members to bond and build trust with each other. At the conclusion of the retreats, however, I felt less pessimistic and realized that it is not only possible for community members to consistently do such work together, but that when they do, they successfully break through their internalized culture of empire and create and sustain the connectedness that empire renders utterly impossible.

I'm not talking about momentary feel-good experiences where everyone holds hands and dances around the world as it were, nor am I talking about everyone agreeing on every issue. I'm talking about the kind of profound, intimate joining that natural cultures of indigenous traditions were able to experience and sustain and which allowed them to survive and thrive. And while circles of community building do not guarantee survival in the face of collapse, they are remarkably effective in facilitating the navigation of collapse.

One caveat regarding dialog circles is that they should never be used as a substitute for addressing practical issues such as paying bills, running a business, or attending to day to day chores. When dialog replaces duty, it becomes quite simply, dysfunctional.

So what actually happens in a circle? To begin answering that question it is important to understand that a community circle must be leaderless. Individuals may take turns facilitating circles, but everyone in the group must be a leader. Facilitation simply means bringing up a topic or restating one that is already on the front burner and making sure that the group adheres to already-agreed-upon ground rules. Such ground rules include a commitment to stay in the group until the issue is resolved or until the group decides to take a break or decides to adjourn until a later time. For purposes of safety, everyone needs to agree to stay in the circle and not flee so that when someone is working on an issue with the group, they are not abandoned by anyone and know that space is being held for them by other group members. In addition, members must commit to maintaining confidentiality and making sure that everything that happens in the circle remains there and is not disclosed to anyone outside the circle.

At all times, the group practices deep listening and compassionate truth-telling. When one person is speaking, the rest of the group listens attentively and stays present with the speaker. Likewise, when one speaks, one does so non-judgmentally using "I" statements, speaking as much as possible from a place of feeling rather than intellect or thinking. Perhaps most importantly, each person is accountable and takes responsibility for his/her part in whatever concerns or complaints he/she verbalizes. Deep listening and processing may involve other factors, but these are some of the most fundamental.

In an article published in 2008 by Sally Erickson and I entitled, **"We Can Survive, But Can We Communicate?"** we offered a detailed explanation of dialog circles which many individuals have found useful for understanding the process and its intention, and I have included the article in its entirety below.

CREATING AND FACILITATING DIALOG CIRCLES

When we think of preparing our minds, bodies, hearts, and living situations for collapse, the focus is often on our individual or household living situations. Equally important is our need to develop a circle of trusting, mutually interdependent relationships. The culture we live in is based on hierarchies of control and influence. Work relationships, kept in place largely by paychecks and ordered by project managers and bosses, are the most common experience most of us have of being part of an organized group. We have little experience outside of those hierarchies. Even more rare in our hyper-independent culture is to depend on others for mutual aid, support and comfort. So, for most people, it likely feels overwhelming to consider how to build a wider circle of people based on mutuality, as part of preparation for the ongoing collapse of basic life support systems.

As daunting as that challenge may seem, consider that individuals in isolation will have a hard, lonely, and extreme challenge if they try to survive the world that will remain when systems collapse with ever-increasing rapidity and intensity. Humans are hard-wired as social beings. Absent the distractions of media and entertainment we will find that we need each other. At the same time, we will discover how emotionally and spiritually wounded we've become as members of the largely bankrupt, and often abusive, culture that empire has created.

Sadly, peoples' experiences of community end all too often in pain and disappointment. Such experiences range from attempts to live in intentional communities to the struggles of serving on church committees or being part of activist organizations. As a whole we are ill-equipped to create cohesive and cooperative groups and then to resolve ongoing issues and conflicts that naturally arise. People often express cynicism, despair and helplessness around the possibility of successfully creating and maintaining a sense of working community within a culture of empire. Clearly, it is critical to acknowledge the need for a sense of real connection, for the ability to work through conflict, and to cooperate in effective and joyful ways with others. Once we have come to terms with the need to do so we can begin to find others who have identified the same need and are ready for the task.

Let's first identify what we are talking about when we talk about community. In this context community does not refer only to individuals or families who own land together or who happen to live in proximity to one another, although proximity will more and more be the rule as fuel becomes scarce and travel is limited. We define community, in this context, to be a congregation of people who have, by the commitment and skills they possess, learned to establish relationships characterized by trust, understanding, mutual respect, and bonding that transcends personality and allows and even embraces differences of background or ideology. Such a group is able to think together effectively and to tap into deep wisdom about challenging issues. They can do this because they trust each other enough to question and suspend the assumptions and core beliefs that limit their insights as individuals. Such a group does not come together, as a therapy group does, for the purpose of healing per se, although insight and healing of isolation, unresolved past conflict, fears, and insecurities often occurs. The purpose of the kind of community we are speaking of is to come together to glean wisdom from listening and speaking with one another and to offer connection, support, comfort, and mutual respect. Such a group of people learns together to find better solutions, wiser actions and more joy together than is possible for them to do as isolated individuals, couples or families.

When defined in this way, the idea of community appeals to most people, even when they doubt their ability to find or create such an experience. But the times demand that we do what we've not believed we are equipped to do. It helps to remember that humans are indeed "hard-wired" for this. Indigenous peoples overall have felt the benefits of inclusion in close-knit social units. It is the wounding of the current culture that has disrupted that hard-wiring, often for many generations, and certainly most seriously in current times. But deep trust and connection is something we need in order to feel fulfilled and secure. Once accepted, the need to build community is simply another task to attend to as the current system unravels.

As tempting as it is to focus only on the logistics of living arrangements, how resources and tasks can be shared, preparation for crisis conditions, and other issues, it is equally important to develop skills to create and maintain authentic connection and to work through conflict. When groups fall apart it is almost always as a result of emotionally charged issues. It is important that people make a commitment to find ways to work with people's emotions, to communicate fully, and to bond. Groups will do well to cultivate skills in listening and truth-telling, because when emotional issues are not consciously addressed and worked through, they often sabotage a community's very existence. At the very least unresolved conflict makes life miserable and drains huge amounts of energy that would better be utilized attending to other needs. Much talk of ecovillages and intentional communities abounds among people preparing for collapse. Evidence that dealing with relationships is essential is the fact that most of these situations devote a significant amount of time to building a workable sense of community.

Conflict is inevitable. A community must develop skills to effectively resolve conflict so that people feel cared-for and respected. Its apparent absence is a red flag signaling the likelihood of dysfunction, of unspoken feelings and truths that need to be told, or of a strict authoritarian hierarchy that keeps conflict as well as individual creativity submerged. Indigenous cultures at their high points skillfully navigated conflict, and in fact probably welcomed it. They evolved creative skills for addressing it compassionately and assertively, with elders, both men and women, who carried those skills and wisdom down through generations. Those of us reared in the hierarchies of empire are not so lucky. Most people don't feel fully adult much less secure enough to be considered real elders. We are having to glean the best we can from older cultures and learn from the most innovative practices that have come from psychology and organizational development to find our way in to creative, cooperative relationship.

Here are some insights that may be useful:

People who have had opportunities to sit in listening/ truth-telling circles often at first feel overwhelmed with the amount of emotional work that needs to be done in order for group members to bond and build trust with each other. This has certainly been our experience. But when people make the commitment and see the process through the difficult stages, they find new optimism. Groups that break through to what Scott Peck called "true community," experience what human beings are capable of. Regular people, with the garden-variety neuroses and the wounding that is typical of most of us educated in public schools and reared in the typically dysfunctional families of empire are surprised at the connection possible. What we realize is that community members are able to consistently do this work together, and that when we do, we successfully dissolve internalized patterns that have been inculcated by empire. What we experience in the place of those old patterns is the joyful connectedness that empire had rendered utterly impossible.

Those who have participated in community-building workshops and other kinds of training in dialogue and human interaction find this is a repeatable experience. People find they are able in this work to include and allow for differences. This experience is akin to the profound, intimate joining that indigenous people experience and sustain, which has allowed them to survive and thrive. Such experiences of mutual respect, understanding and bonding are likely to support individuals and groups in critical ways during time of societal upheaval.

There are principles that underlie effective group interaction. It helps immeasurably to have one or two strong facilitators present who are familiar with the inner terrain a group must travel to develop trust and to transcend differences. The process is rarely smooth. Facilitators are different from what we generally think of as leaders. Facilitators help the group, as a whole, move into shared wisdom. This is very different from a group that accepts and follows the wisdom or philosophy of a charismatic leader or the dictates of an authoritarian leader. Rather, this kind of community may be said to be "a group of leaders." Each person is regarded as someone who brings a unique set of gifts, experiences, skills, and insights. Strong facilitators help empower individuals to share those individual qualities for the greater good of the group.

Key to building this kind of community experience is the practice of compassionate listening and truth-telling. When one person speaks, the rest of the group listens attentively and stays present with both heart and mind. Speakers "speak from the heart" and speak when truly moved to speak rather than compulsively or in reaction. Another key is that each person learns to take responsibility for his/her part in whatever concerns or complaints he/she identifies. This requires each individual to examine his/her own assumptions and core beliefs and patterns, and to risk sharing those with the group so that they can be examined and understood.

What follows are some "Principles Of Dialogue" that Sally Erickson has synthesized from group development theory, Scott Peck's model of community building, and David Bohm's explorations of formal dialogue practice.

Principles of Dialogue

<u>Coming Together</u>

1) We agree to identify and suspend assumptions and core beliefs. Suspending doesn't mean eliminating. It means holding them aside so as to be able to listen more deeply to another's experience, knowledge, insight. It means being willing to allow beliefs and assumptions to shift as the conversation reveals greater insight and understanding.

2) Examples of three kinds of assumptions/core beliefs:

**Factual: I assume energy can/cannot be created by hydrogen.

**Personal: I assume I am/am not personally responsible for saving the world. I assume I am/am not valued by those around me.

** Spiritual/philosophical: I assume that the material world as mapped by Newtonian physics, chemistry, biology, is/isn't all there is to reality. I assume human beings are/are not the pinnacle of evolution. I assume there is/is not a power greater than the human ego.

What happens when we suspend our assumptions and question core beliefs? We are likely to experience initial anxiety. As we sit through that anxiety, habitual ways of thinking, feeling, and being soften and we find new possibilities. For example, if we usually talk a lot in a group, we begin to listen more. If we usually don't talk, we find the courage to speak when moved to do so. If we tend to stay in our intellect, we notice and identify our feelings and are more aware of our bodies. If we tend to be largely in our feelings and body, we begin to use the mind and insight more. Long-held beliefs and assumptions are revised or abandoned in the light of new information and insight. Group wisdom emerges that is greater than the sum of the collected individual's knowledge.

3) We agree to come together as colleagues. While individuals are not necessarily equal in specific knowledge or skills it is important to regard ourselves and each other as equal in value. Each person possesses unique abilities in a variety of arenas that are important to the community: insight, ability to listen and

be present, intuitive gifts, dreams, clarity, connection to the natural world, as well as factual knowledge, skills, etc. When we come together as colleagues we make a commitment to notice the tendency to regard ourselves, and others, as either higher or lower. And we agree that when we notice that tendency we will work to open to find the unique value of others and ourselves in cooperation.

Group Norms and Standards

** We agree to confidentiality. To increase a sense of safety, it is important that members who come together to do this kind of work commit to maintaining confidentiality. We agree that what is shared in the circle will not be shared outside of this circle in any way that would violate the confidentiality of the members of the circle. One's own experience can be shared outside, but names, other's personal stories or what actually occurs during the circle will not be shared or gossiped about.

**We agree to show up and be present. This commitment helps members feel some degree of emotional safety. Having been raised in empire we almost all have felt abandoned when we expressed vulnerability and were trying to be genuine and honest. When everyone agrees to stay in the circle and not flee in the face of conflict or discomfort, "the space is held." As vulnerability surfaces and conflicts are confronted, the result is that everyone feels safer and more willing to risk telling their truth. Trust is built incrementally but undeniably when people "hang in there" for the long haul.

**We agree to take the time that is necessary to do the work. It has been the experience of many groups that it takes a minimum of two full days, or 16 hours of interaction, for a group to begin to establish the kind of trust and openness that yields the fruit of real dialogue and bonding. It is generally wise to schedule more than that number of hours in order for a group to really coalesce and begin to learn to work well together. It is important that all participants agree to be present for all sessions. Occasionally exceptions can be made, but generally people who miss out on the work the group does together will not develop the same level of trust.

This is a critical point to note. All too often, just as a group is about to break through into a new and more profound level of functioning, interactions will get very challenging. People will get discouraged and want to quit or take a break to do something else. This is the part of the process Scott Peck called "emptiness," and it *is* challenging to get through. It is at this point that a strong facilitator can be especially helpful in giving the group confidence, in "holding the space." By his or her presence the group will find the courage to keep working rather than to flee into some other activity.

**We agree that no one is required to speak, only to work to be fully present. Since many people feel intimidated about speaking in large groups this agreement encourages people to be involved who might not otherwise participate. Often the attentive presence of very quiet people will add immeasurably to the experience. And often their verbal contributions will be spot-on when they are made. Because of the nature of the work and the need to be mentally clear and emotionally available to the experience, participants agree not to use mood altering substances including alcohol for the duration of the days that the group is engaged in the work.

**We agree to be mindful and to resist "sub-grouping." There is a natural temptation to talk in pairs or in small groups during meals and breaks about charged feelings that arise as a result of the work of the circle. It is very important to bring those expressions of feelings into the "container" of the group or there may be a tendency for factions to develop. While the tendency to "process" outside of the group is understandable if feelings and insights and challenges are not shared within the group, its power is diluted, and the process of building trust will be prolonged. Withholding unresolved feelings and conflict and factioning as a result can ultimately sabotage the work.

Interventions In The Process:

Since the facilitator is there largely as a guide and elder, but not as a sole leader, others are encouraged to intervene in the process when they begin to feel stuck or frustrated with the way things are going. Participants are encouraged to put words to their feelings of frustration and then to request that the group consider reflecting on the process and work to shift it. Anyone can make the following requests to help the group work more effectively.

1) Ask for a minute of silence.

2) Ask for people to identify, talk about, and suspend their assumptions around an issue.

3) Ask for each person to hold the question: "What is it in *me* that is keeping us from going deeper?"

4) Ask the group to try to "speak from the heart."

5) Ask for each person to question: "Am I taking full responsibility for *my* part in whatever is going on right now?"

6) Ask the question: "Is frustration present and if so, what is the nature of the frustration?"

7) Ask the question: "Is there something we are not talking about, and if so, what are the assumptions we hold that keep us from talking about it?"

With every passing day it becomes clearer to us that as civilization continues to self-destruct, we need to discern how we prefer to spend our time and energy, and with whom. What we least want to do is mimic the culture of empire by limiting our focus to logistics, thereby losing sight of our deep humanity. We know that we cannot survive alone. Even if we have learned every physical survival skill imaginable, we still need our fellow human earthlings in order to navigate collapse. Moreover, if I and my companions in collapse cannot deeply listen to each other and speak our truths with compassion, even if we survive, it will be within an internally vacuous emotional domain that would render survival nothing less than absurd.

A William Stafford poem "A Ritual To Be Read To Each Other" illumines our need for deep connection:

If you don't know the kind of person I am
and I don't know the kind of person you are
a pattern that others made may prevail in the world
and following the wrong god home we may miss our star.
For there is many a small betrayal in the mind,
a shrug that lets the fragile sequence break
sending with shouts the horrible errors of childhood
storming out to play through the broken dyke.
And as elephants parade holding each elephant's tail,
but if one wanders the circus won't find the park,
I call it cruel and maybe the root of all cruelty
to know what occurs but not recognize the fact.
And so I appeal to a voice, to something shadowy,
a remote important region in all who talk:
though we could fool each other, we should consider-lest
  the parade of our mutual life get lost in the dark.
For it is important that awake people be awake,
or a breaking line may discourage them back to sleep;
the signals we give-yes or no, or maybe-should be clear:
  the darkness around us is deep.

Stafford reminds us of how important it is to know each other in a world where the culture of empire and its "patterns that others have made" may cause us to follow the wrong god home. Not only must we know each other, but we must, like elephants connected by trunk and tail, hang onto each other in order to find our way. We could fool each other, but we dare not because if we do, we may get lost. Therefore, it is imperative that we be awake and that we be transparent with each other because the darkness around us is so deep; our commitment to community is essential in navigating that darkness.

The rewards of investing our time and vital energy into our community are infinite and succinctly captured in the words of author and psychotherapist, Thomas Moore in *The Re-Enchantment Of Everyday Life:*

> *When we all, leaders and participants in community, discover the sheer joy of creating a way of life that serves families, ennobles work, and fosters genuine communal spirit, then we will begin to touch upon the sacredness that lies in the simple word polis, which is not just a city defined in square miles, income, or population, but a spirit that arises when people live together creatively.*

Other communication models include Marshall Rosenberg's Non-Violent Communication process[38], Peter Senge's leadership training materials and workshops[39], Tej Steiner's Heart Circle[40] work, Council Training at the Ojai Foundation[41], as well as Scott Peck's work[42]. Resources in one's home locality ought to be considered as well.

A combination of modalities may be useful, but what is just as important as the method is the community's commitment to the process of healing the wounds of empire both internally and as they manifest in our relationships with each other. As we move out of the disintegrated structures of the culture of empire there is a tremendous opportunity to move into integrated and joy filled structures of relationship, inner and outer, with ourselves, one another and the whole community of life. Relationships that bring comfort and joy will be a mainstay as we sail through these most difficult times ahead. In addition, dialog circle work can facilitate our finding a greater group wisdom about how to navigate these times than we can find on our own.

REFLECTION

**What have been my experiences, both positive, negative, and in between, with community? Be specific.

**What fears do I have about participating in community? Be as specific as possible.

**What am I afraid to reveal to my community, whatever or whoever my community is? Why am I afraid of this? What am I afraid I might find out about the other members of my community? Why?

**What fears do I have about sitting in a circle and engaging in deep listening and deep truth-telling?

**How do I envision healthy, loving, safe community? Who would I like to have in my community?

NOTES

NOTES

Chapter 13–Raising the Next Generation As the World Ends

Wind whines and whines the shingle,
The crazy pierstakes groan;
A senile sea numbers each single
Slimeslivered stone.

From whining wind and colder
Grey sea I wrap him warm
And touch his trembling fineboned shoulder
And boyish arm.

Around us fear, descending
Darkness of fear above
And in my heart how deep unending
Ache of love!

~James Joyce~

The collapse of civilization will be easy for no one, but especially for parents who are aware of collapse but who, upon becoming parents, probably had little inkling that their children would not have an easier life than they had. All parents desire a better world for their children and usually sacrifice their own needs for nearly two decades to that end. For some parents, the likelihood that if their children survive collapse, they will be forced to live in a society in tatters amid danger, deprivation, and exceedingly formidable challenges is a prospect which they simply cannot allow themselves to contemplate. Unless the human race becomes extinct, it will be, after all, the next two or three generations of humans on earth who will be forced to pay the price of the previous dozens of generations who were unwilling to notice the despicable legacy being left to their children—or who, although aware of it, believed themselves powerless to prevent it.

A 2007 article in the *London Sunday Times Online* entitled "Because We're Worth It"[43] states that, "The baby-boomers' culture of hedonistic consumerism has left their offspring with the crumbs from their table. And 65% of them say their children's lives will be worse than their own." The article concludes that one of the tragic realities of the boomer generation is that, "At the heart of this short-termism is the deep cultural truth that boomers have lost the old, philanthropic view of the future. Something happened to my generation on the way to the 6000 square-foot house, the plasma TV screen, the SUV fleet, the hot tub, and the multiple vacations in Hawaii—all financed with debt that mortgaged their children's and grandchildren's futures."

Many parents have told me that until recently, as the global economic meltdown has intensified, their older children refused to believe in the reality of collapse and wondered about their parents' sanity when mom or dad made preparations for surviving post-civilization. Other parents found themselves engaged in struggles with older children or teens who, bombarded with peer pressure, balked at powering down, shopping at thrift stores, and generally living a more sustainable lifestyle. Still other parents have been holding fascinating conversations with their children about collapse, depending of course on the age of the children, and have reported finding pleasure in doing so.

Perhaps one of the most challenging topics when parents dialog with children about collapse is why parents are choosing to live more simply and why certain items, products, or activities are no longer affordable or desirable in a household committed to sustainable living. If parents have been teaching and modeling sustainability for some time, this may be less of an issue than in families where parents are making rapid changes in that direction. Depending on the community in which the family resides, pressure to do, buy, and have in the same manner as one's peers may or may not be a momentous issue.

At this writing, children are beginning to watch parents become unemployed, file for bankruptcy, experience the entire family losing the house to foreclosure, or at worst, become homeless. If there has been no preparation for this, these children are getting a painful crash course in the limits of economic growth and consumption.

Before the current economic crisis, some collapse-aware parents were dialoging with children about values, family finances, what is affordable and what is not, and for those families, the less challenging the maintenance of a lifestyle of simplicity is likely to be. Perhaps children in those families who were old enough saw "The End Of Suburbia", "Maxed Out", "In Debt We Trust", and "What A Way To Go" with their parents and may have also watched the 2007 short documentary "The Story Of Stuff" which is perfect for sharing with middle school, teens, or college students. Animated and narrated by a young woman of the millennium generation, the film uses humor, statistics, and a tone of "C'mon guys, let's consume less and care about the planet more" to convey its powerful message.

In the fall of 2007, I interviewed Lisa McCrory and Carl Russell[44] of rural Vermont who operate *Earthwise Farm and Forest* which teaches a variety of skills for sustainable living, including the use of draft animals in raising organic crops. When I asked them about the extent to which civilization influences their children, Carl responded by saying that "It should be understood that we have cell phones, laptops, CD players, DVD/VCR-TV, and game-boys. What our off-grid sustainable lifestyle does, is put these things into a subclass of luxury and leisure. We teach our kids the language of our modern culture because it is necessary for them to function within their community. We do not shun modern culture, or try to hide from it, but we strive to teach our children the language of the Earth, about the spiritual and physical truths of human life on planet Earth. We entertain acquaintances as we process chickens, as many people seek our guidance with the skills of slaughtering and butchering their own animals for food. One day as I was removing entrails, our 5-year old son cheered, 'Chicken Livers'! Our visitor turned to me with a look of astonishment. 'How many modern 5-yearolds know enough about intestines to know where the liver is, and how many of them would be excited about eating it, especially after seeing where it comes from?' "

Lisa then added: "Although we do have all the things that Carl has listed above, we *do not* have access to public or cable television, so are not heavily influenced by commercial advertising, the constant marketing targeted towards children, and the media-driven 'news' that to me is about 20% news and 80% questionable. We watch movies that we choose when it meets our schedule. We also home school our children which we feel has been very rewarding for our children and for ourselves (ages 3, 5 and 10). That said, our 10 year-old is going to the public school for some electives (music, art, math, soccer, band). I think that our kids are very in touch with where their food comes from and what it takes to make that happen. We went to eat at a friend's house not too long ago and our 3 year-old started asking questions about the food on the table; 'Did you kill this chicken?' and other questions like that. Our 5 year-old was amazed to find out that this family did not have any farm animals and said 'You mean you don't even have one cow?' Hilarious what comes out of the mouths of babes!"

In families such as Lisa and Carl's, collapse preparation will not be a daunting issue because their children have been living sustainably all their lives. It appears that they would not want or need to live any other way.

Home schooling is invaluable for parents such as Lisa and Carl and others who are committed to teaching their children the values of simplicity and earth-connection. It also offers many opportunities for building and deepening community among parents and children. In addition, home schooling is an ideal venue for teaching children about local economies and taking short field trips to small businesses in the community where they can converse with merchants, farmers, bankers, and other professionals about the realities of local commerce.

Moreover, as collapse intensifies and public schools increasingly cease to exist, home schooling is likely to become the *only* schooling available to children and adolescents.

In recent years, more parents have been rejecting the *modus operandi* of public schools where standardized testing, motivated by the desire to produce standardized children, results in graduates who are woefully uninformed and incapable of thinking critically. As collapse exacerbates and the consequences of Peak Oil curtail the most fundamental activities of civilization, the demise of public education is all but guaranteed. All the more reason that parents be trained to home school not only their children but those of their friends and neighbors. Included in every home school curriculum should be the teaching of fundamental survival skills as well as a basic, "three-R's" education.

The stress of collapse is already impacting many families brutally, especially those who are unable to discern its deeper realities and navigate its losses. As in all stressful situations involving parents and children, anxious parents may want only to be comforted themselves, but will wisely choose instead to comfort their children. However, parents must get their own needs met; therefore, living cooperatively with many families sharing in childcare, home schooling, and other community tasks will become increasingly desirable and necessary. Thus, another opportunity that collapse may offer, in my opinion, is a community setting where we finally discover that not only does it take a village to raise a child, but that for the first time in our lives, the village is actually there!

Older children and teenagers should be encouraged to form their own circles of listening and truth-telling and also be included in adult circles if they desire. We must welcome their wisdom, for they are the ones who are likely to abide when the elders have departed.

The progression of collapse is likely to result in larger extended families sharing one living space. As college and career options dwindle and unemployment and bankruptcies increase, single children who have left home may return to live with parents or if married, may bring their families with them. Likewise, grown children will unequivocally be forced to care for aging parents in their own homes. The greater the extent to which family members have consciously prepared for collapse, the better they and their families are likely to fare in navigating it.

In situations where families are sharing land and living space, the opportunity for developing cottage industries that offer necessities to the larger community may proliferate. Selling or bartering products and services such as food, natural healing, tutoring, transportation, mechanical repair, home maintenance, and other goods and services is likely to become widespread, and families can create enterprises that sustain themselves and their community. For this reason, it is important that parents assist children in learning the same survival skills frequently mentioned in this book that they are themselves learning.

Massive unemployment is making glaringly obvious the reality that many of the careers for which parents and children have sacrificed to prepare for by going to college, will not exist a decade from now. Therefore, as parents and children discuss these realities, children should be strongly encouraged to listen to their hearts and minds regarding the skills they need in order to make a livelihood and survive a devastating economic depression from which this nation and the world are unlikely to recover.

A crumbling world could offer as much opportunity for healing among family members as many years of family therapy. When all family members fully understand the reality of collapse and that each of their lives depends on the well being of every other family member, myriad wounds are guaranteed to surface, but so might moments of deep listening and truth-telling that could facilitate forgiveness and transformation.

REFLECTION

**If you have children of any age, take some time to journal about your concerns for them in a collapsing world. At this point in their lives, how can you most effectively assist them in preparing for collapse? Be specific.

**How can you assist your children in learning survival skills? What resources do they already possess for navigating collapse?

**What causes you to tremble about your children's future? What brings you comfort when you contemplate their future?

**Perhaps you have done nothing directly to rob your children and grandchildren of a prosperous future, but your generation has, as have all generations older than your children. Have you apologized to them for this reality? If so, what happened when you did? If you have not, consider doing so and then journal about what happened.

**Can you talk with your children about collapse? If you have done so, what has been the dialog and how has it proceeded? If you are unable to discuss collapse with your children, how does that feel? Spend some time journaling about that.

**Can you discuss collapse with other parents? If so, what have those discussions been like? If not, what has that been like? Take time to journal about this.

**If your parents are living, are you able to discuss collapse with them? If not, how does that feel? If you are able to talk about collapse with your parents, what is that like for you?

**What are some family wounds that may be surfacing as a result of the current manifestations of collapse? Take some time to journal about your feelings with regard to those wounds. What support do you need that you do not now have?

**Have you experienced healing with family members in your discussions with them about collapse? Has any aspect of your conversations brought you closer together?

NOTES

Notes

CHAPTER 14–MIRTH-MAKING AMID CRUMBLING AND CHAOS

We are really, really fucked, and life is really, really good.

~Derrick Jensen~

I will be the first to admit that while extremely rewarding, writing this book has not been "fun". Every chapter has been filled or tinged with heaviness, but from the moment I pulled myself out of denial regarding collapse and made a commitment to tell the truth about it, I have taken on the role of the bearer of bad news. Articulating the reality of the crumbling of civilization does not promote popularity nor endear oneself to the masses who insist on hearing only good news and who upon hearing the opposite, seem insistent on clutching to their chests, teddy bears of "hope."

Nevertheless, as my students who endure an entire semester of my classes and often sign up for more are fond of telling me, they appreciate how my sense of humor helps to balance the bad news I deliver. Perhaps that's because I enthusiastically agree with the words of Derrick Jensen above. Even in the face of the collapse of civilization and the incomprehensible suffering that will entail, life is really, really good, and I have no doubt that even in the throes of collapse, we will find moments of joy, fun, and play. In fact, we may savor them more than we do now because they will temper the impact of the darkness around us and because they are likely to be fewer and farther between than they are now. In fact, a unique and profound quality of joy may naturally flow as people become less attached to civilization and more attached to nature and the kind of simplicity for which they have opted or which has been foisted upon them.

Undoubtedly, we will have ample reasons and opportunities to celebrate even in the midst of very hard work and much anxiety. Fortunately, the human heart does not need to learn how to laugh—except that maybe some hearts do, especially in the context of the great unraveling of civilization. What is more, we must be aware of the likelihood that our merriment may occur against the backdrop of untold suffering around the world. Survivor guilt may inhibit us; numerous paradoxes comprised of countless polarities may cry out for resolution. But as we noted in Chapter 4, practicing gratitude will sooner or later open the door to joy.

I greatly admire James Howard Kunstler for incorporating in his dark novel about a post-collapse world, moments of joy and celebration, as residents of his "world made by hand" were able to organize festivities which included sumptuous organic dishes, homemade wines and beer, music played on instruments they had made with salvaged materials, and lighting and decorations ingeniously improvised in an off-the-grid society.

When deciding to sell or give away items we no longer need, we should not be too quick to part with musical instruments. Civilization has created such dependency on high-tech entertainment that its citizens often feel unable to have fun without it. Therefore, items that can facilitate mental and emotional distance from the stress of our lives and which do not require electrical power for use should be prized and preserved. Again, I'm reminded of the movie "Playing For Time" and the comfort and meaning instruments gave to the members of Fania Fenalon's musical group in Auschwitz.

As a drummer of the African djembe my hope is that many drums and drummers will inhabit the days and nights of my community during collapse. Indigenous peoples have always perceived the drum beat as an "echo" of the heart beat. Over a decade ago I began attending workshops where powerful stories were told by a storyteller who drummed while telling the story. I was profoundly moved and my consciousness altered as my entire body absorbed not only the story, but the drum's compelling rhythms. In fact, I was so moved that I purchased a drum and took a few lessons from an experienced drummer. Before long I was able to speak the words of stories while playing the drum. In workshops over the years I have told many stories while drumming, and without exception, I have remained continually in awe of the power of this process for the audience and for me.

Yet even without story, drumming—particularly group drumming, connects us with what indigenous people sometimes call "the other world", and the easiest place to begin accessing that world is within ourselves. Drumming can deliver us to the transcendent within—to a place beyond the dilemmas of the physical, mundane world where something more mysterious and majestic informs the ubiquitous crumbling of civilization. It can expose chasms of sorrow or evoke rivers of laughter; it can exhilarate as well as soothe. I pray for drums in the darkening days of collapse and for many instruments, including the most primal—the human voice—with which we can make glorious music. I pray also for dancing and the liberation of our bodies for gyrating, writhing, undulating, and convulsing with laughter, release, abandon and reckless abandonment of civilization. A lovely poem entitled "Advice" by Bill Holm reminds us of why we must dance:

> *Someone dancing inside us*
> *Learned only a few steps:*
> *The "Do-You-Work" in 4/4 time*
> *The "What-Do-You-Expect" waltz.*
> *He hasn't noticed yet the woman*
> *Standing away from the lamp,*
> *The one with black eyes*
> *Who knows the rhumba,*
> *And strange steps in jumpy rhythms*
> *From the mountains in Bulgaria.*
> *If they dance together,*
> *Something unexpected will happen.*
> *If they don't, the next world*
> *Will be a lot like this one.*

Obviously, as we celebrate and play, suffering will exist around us and within our own communities, families, and relationships. Some will experience survivor guilt for having fun while the rest of the world is literally or symbolically burning, and unless we have learned to hold a variety of paradoxes in our bodies and souls, external conditions could inhibit us from salutary moments of joy. If we are able to hold the paradoxes, we may feel a bit like blood-soaked doctors and nurses in hospital operating rooms who play rock music and laugh during brain or bypass surgeries.

One of the most powerful tools for assisting us in holding paradoxes is practicing gratitude because the moment we are able to feel gratitude for anything, we are declaring that regardless of how much darkness may permeate our lives, we have not become totally engulfed by it. Moreover, as we continue to engage with the question "Who am I in the midst of collapse and the suffering it entails?" we are confronted with the reality that some part of us—miniscule as it may feel, has the capacity for experiencing joy. No matter how much guilt we might experience for having fun or feeling playful, the suffering of the world around us does not change the reality that some part of us is capable of laughter.

Another aspect of preparing for collapse might be the cultivation of celebratory activities which do not require electrical power or driving somewhere to entertain or be entertained. As in earlier times, communities might celebrate the completion of spring planting, fall harvest, equinoxes and solstices, as well as celebrating rites of passage for young people who have completed initiation ceremonies or older individuals at the end of rituals marking their transition to elderhood. Significant nature events can be celebrated such as the first snowfall, a long-awaited rain in a dry season, birds returning from winter migrations, a successful hunting expedition which provides food for the community, and the outbreak of resplendent autumn foliage. Females during their menstrual cycle may wish to reconstruct the observance of the Native American moon lodge and gather in it during the week of their "moon event" for relaxing, telling stories, dreaming and recounting dreams to each other, and of course, laughing and playing.

What is certain is that collapse will compel us to simplify everything we do, including how we entertain ourselves and where we find pleasure and amusement. The more we have come to rely on technologically intensive entertainment, the more frustrated and deprived we are likely to feel in a world where it is no longer available.

Of course, the choice is always ours to succumb to despair and simply give up. The repercussions of collapse may feel intolerable, and depression, suicide, ennui, or bitterness may overwhelm us as they are certain to overwhelm incalculable numbers of individuals. (A later chapter in this book will address issues of emotional health and emotional exhaustion.) And while laughter nourishes our bodies and souls, it also causes us to be vulnerable, and we may fear that relaxing for just a moment may cause the walls of our defenses to become more permeable. Yet our despair may be tempered if we can practice humor alongside gratitude.

Joy and gratitude are inseparable partners as David Stendl-Rast explains:

> *Joy is that extraordinary happiness that is independent of what happens to us....The root of joy is gratefulness. We tend to misunderstand the link between joy and gratefulness. We notice that joyful people are grateful and suppose that they are grateful for their joy. But the reverse is true: Their joy springs from gratefulness. If one has all the good luck in the world but takes it for granted, it will not give one joy. Yet even bad luck will give joy to those who manage to be grateful for it....It is not joy that makes us grateful; it is gratitude that makes us joyful.*

Carolyn Baker, Ph.D.

REFLECTION

**In what ways do you currently experience joy, fun, and play? To what extent are those experiences dependent on abundant hydrocarbon energy or technology?

**In what ways do you imagine yourself experiencing joy and play that do not rely on hydrocarbon energy?

**When you are anxious or worried, is it difficult for you to take a break from that stress, and play? Do you imagine yourself feeling guilty for having fun when the world around you is suffering?

**What musical instruments do you play? Even if you do not play a musical instrument, what instruments would you like to have in your possession when access to traditional forms of energy are limited or inaccessible?

**What experiences have you had of holding humor, joy, or play alongside sorrow or adversity?

**What are your experiences of connecting gratitude with joy?

NOTES

NOTES

CHAPTER 15–TELLING ANCIENT STORIES AND WRITING NEW ONES

Stories set the inner life into motion....Story greases the hoists and pulleys, it causes adrenaline to surge, shows us the way out, down, or up, and for our trouble, cuts for us fine wide doors in previously blank walls, openings that lead to the dreamland, that lead to love and learning, that lead us back to our own real lives....

~Clarissa Pinkola Estes~

Myth gives a person the sense of living in a meaningful story, the feeling that one's life makes sense and has value, and these sensations are the basis for self-confidence and stability, purpose and poise. Without myth, life has to be proven valuable every day and is lived from profound anxiety; but with the awareness that one's life is grounded in eternal stories and motifs, one's own personal story begins to feel enchanted, and this feeling gives rise to a love of one's own life that is the cure for narcissism, insecurity, and self-doubt.

~Thomas Moore, *The Re-Enchantment Of Everyday Life*~

For more than a decade I have been a professor of history. It is no secret to anyone who knows me or who has read my book, *U.S. History Uncensored: What Your High School Textbook Didn't Tell You* that I strongly object to the manner in which history is taught in most colleges and secondary schools—if it is taught at all. As I listen to my students who share their experiences of taking history classes in high school, I have been appalled at the dearth of real history instruction that actually occurs in public schools. Students consistently report that they were bored to tears during high school history, sometimes to such an extent that they cannot even recall *taking* a history course in high school.

I have discovered that most human beings love to hear stories, and I believe that the capacity to hear and resonate with story is part of our DNA and our ancient memory. I've conducted my own experiment around this in the classroom. When I present a lecture on some aspect of history, I can count on some of my students losing interest and in some cases, falling asleep. However, when I cease lecturing and begin telling a story, they come back to life. Eyes stop drooping, heads stop nodding, and suddenly, they are with me. For this reason, I prefer teaching "his-tory" or "her-story" by simply telling the story.

Stories shape our lives from the moment we are born. Some may argue that an infant cannot comprehend a story, but I must ask, what does it mean to "comprehend"? On some level of cellular, ancient memory the infant "hears" and "comprehends" the story. As she matures she hears it in a different way, on a different level.

David Abrams in his article "Earth Stories", reminds us that the unique linguistic and intellectual capacity of humans did not evolve through computers or even the written word, but in relation to orally-told stories which were being communicated long before words were written down. Abrams emphasizes that:

> *Spoken stories were the living encyclopedias of our oral ancestors, dynamic and lyrical compendia of practical knowledge. Oral tales told on special occasions carried the secrets of how to orient in the local cosmos. Hidden in the magic adventures of their characters was precise information regarding which plants were good to eat and which were poisonous, and how to prepare certain herbs to heal cramps, or sleeplessness, or a fever. The stories carried instructions about how to construct a winter shelter, and what to do during a drought, and - more generally - how to live well in this land without destroying the land's wild vitality.*

Because of our innate capacity to hear and resonate with stories, we absorb them like a sponge, and therefore, the kinds of stories we hear profoundly influence our thoughts, feelings, and values. In the culture of civilization we hear hundreds of stories before we enter kindergarten, and those stories inculcate civilization's paradigm. We hear stories that teach us that we are separate from nature and the non-human world, that humans are superior to animals, that humans are innately flawed or that humans are sinful, that the United States is the greatest country on earth, that heterosexual orientation is preferable to a gay, lesbian, bisexual, or transgendered orientation. We are told subtly or blatantly that white is the most desirable skin color and that making money is the most valid motivation for living our lives on planet earth. Boys hear stories of heroes and conquest; girls hear stories of beautiful women and girls who are nice "like sugar and spice."

But the hubris of humanity has led us to believe that stories have only been created and told by people whereas in the words of Abrams, "The stories themselves were carried by the surrounding earth. The local landscape was alive with stories! Traveling through the terrain, one felt teachings and tellings sprouting from every nook and knoll, lurking under the rocks and waiting to swoop down from the trees."

One story which begs to be retold in our time relates to the mythology of the ancient Mayan people. According to their cosmology the winter solstice in the year 2012 will bring a rare astronomical alignment of the sun with the center of our galaxy—an alignment that only happens once very 26,000 years. The Mayans believed this event would mark the end of a cycle, but it is important not to literalize the story of this event and miss the symbolism of it which the Mayans held would be far more momentous than the event itself. A fascinating 2007 documentary "2012: The Odyssey", focuses on the Mayan calendar and interviews a number of researchers who elucidate its spiritual and cosmological implications.

One of these is John Major Jenkins, author of *Maya Cosmogenesis* and *Journey To The Mayan Underworld*, who explains that for the Mayans, the 2012 solstice symbolizes the death of humanity's out-of-control ego and the return of humanity's ego to the divine, eternal self. Not only is the alignment symbolic in terms of the ego and the sacred self but also with respect to masculine and feminine energies. The Mayans believed that the Milky Way is the Great Mother, and her alignment with Father Sun on December 21, 2012 symbolizes unprecedented integration and balancing of feminine and masculine energies within humans.

Numerous other myths relevant to this time period have endured throughout human history. In fact, almost all spiritual traditions include some sort of endtime prophecy. This is particularly significant in our time as fundamentalist Christianity endeavors to convince the world that its notions of the endtime are unique. So-called "prophecy experts" such as Hal Lindsey in his *Late Great Planet Earth* and Tim LaHaye and Jerry Jenkins in the their *Left Behind* series have transmitted fear and awe into the hearts of millions of readers and no doubt amassed millions of dollars from book sales and speaking engagements, yet thousands of years before Christ, ancient peoples were delivering endtime prophecies.

Likewise, centuries before fundamentalist Christians began prognosticating their version of the endtime in North America, Native American elders were teaching their children about the a collapsing world. A number of tribes on this continent are reminding us of their ancient prophecies such as this one from the Iroquois[45]:

> *It's prophesied in our Instructions that the end of the world will be near when the trees start dying from the tops down. That's what the maples are doing today. Our Instructions say the time will come when there will be no corn, when nothing will grow in the garden, when water will be filthy and unfit to drink.*
>
> *Then a great monster will rise up from the water and destroy mankind. One of the names of that monster is "the sickness that eats you up inside" like diabetes or cancer or AIDS. Maybe AIDS is the monster. It's coming. It's already here.*
>
> *Our prophet Handsome Lake told of it in the 1700s. He saw Four Beings, like four angels, coming from the Four Directions. They told him*

what would happen, how there would be diseases we'd never heard of before. You will see many tears in this country. Then a great wind will come, a wind that will make a hurricane seem like a whisper. It will cleanse the earth and return it to its original state. That will be the punishment for what we've done to the Creation.

One notable story which integrates Christian and pagan mythology is the Cross of Hendaye in France containing a number of symbols found on the Mayan calendar. The cross was built by alchemists who were essentially European shamans of the Middle Ages because although they were officially Christian, they studied and utilized pagan mythology and symbolism. Jay Weidner, featured in "2012: The Odyssey" and author of The *Mysteries of the Great Cross of Hendaye* emphasizes that as we analyze the symbolism of the Hendaye cross, we see that it points to an upheaval or collapse which will occur for the purpose of transforming human consciousness. Weidner emphasizes that *those who open to the transformation will have an easier time than those who resist.*

Moira Timms in "2012: The Odyssey" an Egyptologist and author of *Beyond Prophecies and Predictions: Everyone's Guide to the Coming Changes* notes that at the end of cycles, everything comes together—the good, the bad, and everything in between because it is a period of great sorting out—a global initiation that will bring out the best in us as well as the worst. The initiatory process typically involves a shamanic dismemberment in which the ego is shattered, deluged, shaken to its core, so that, as stated above by John Major Jenkins, the ego can return to the divine, eternal self.

With respect to the Mayan myth of alignment in 2012, it is crucial that we not assume that an apocalyptic event will happen on the solstice of that year. What all of the above researchers emphasize is that literally speaking, December 21, 2012 will be a *non*-event because the deeper story of alignment has already begun and is happening now.

I hasten to add that although the documentary "2012: The Odyssey" offers a plethora of possibilities for transformation with respect to civilization's collapse, it does not promise deliverance from the painful consequences of it. In fact, its participants present dire warnings regarding the extinction of the human race and the entire planet. For example, New Mexico author, researcher, and spiritual teacher, Gregg Braden, states that the planet's "global currency" is earth, wind, fire, and air, and that because we are using up that currency on a "planetary credit card", the bills are coming due. We have been "buying now", says Braden, but we will inevitably "pay later." He also emphasizes that we have a probable future, but also a possible future and that the choice will always be ours to intervene in the probable with what is possible.

Psychologist and medical anthropologist, Alberto Villoldo states in the documentary that we must break free of our collective, cultural trance. We live on a planet that can support one billion people riding bicycles, not six billion people who each want three cars and a six-thousand square-foot house, and we will eventually be forced to come to grips with not only our own limits and mortality, but the extinction of the human race.

However, a new story is waiting to be written and drummed. It goes something like this:

> *Once upon a time in a time before time, in the timeless now and approaching a timeless future, humans awakened to feel in the cells of their bodies that within those cells were the very elements of the earth. Feeling the soil, the rivers, the breeze, and the sunlight in their veins, they realized that they had been dreaming; in fact, for several billion years, they had been dreaming a very bad dream—you could even call it a nightmare. They dreamed that they were separate from the earth, superior to it, commanding and dominating it. Soil, rivers, breezes and sunlight were "things" to be managed and allocated, not to feel and savor.*

> *As they awakened, they tasted, smelled, looked upon, listened to, and bathed in the soil until they felt themselves at one with it—until their skin, eyes, hands, and tongues became the soil*

itself. They remembered that the word "human" comes from the Latin word "humus" or earth. No longer did they experience their bodies as comprised of billions of cells but rather as a collection of billions of grains of soil.

And because they felt, smelled, tasted, sounded, and looked like the earth, they felt in every grain of soil/every cell of the body, their union with every living thing on the earth. They did not remain walking upright only, but sometimes crawled on their bellies on the earth, sometimes sat in trees, sometimes writhed in rivers, sometimes held conversations with deer, and all the while, delighted in every breath of air that passed in and out of their lungs.

No longer did they desire superiority over other creatures, nor did they any longer even comprehend what that meant. How could they dominate that which they are, that from which they have never been disconnected? Nor did they feel it necessary to compete with other humans, for they knew that there was enough of everything for everyone, and they especially knew that there was enough love, enough time, enough food, enough beauty, enough joy, enough of everything they could possibly need. And so in the same way that they felt the grains of soil as their cells, they felt the cells of other humans as their own cells.

For this reason, it was no longer necessary to compete, to acquire more of anything, for they realized there was nothing they needed that was not already theirs. Thus the notions of "growth", "expansion", and "conquest" felt superfluous, and absurd to them.

Instead of being motivated by getting, grasping, and hoarding as they had been in their former nightmare, they were now motivated by giving. They desired only to give to other creatures, to care for plants, animals, and other humans as if they were caring for their own bodies because, in fact, as they cared for all other life forms, they **were** *caring for their own bodies.*

Humans now wanted not to take from their children but to give them more than they could ever use for themselves. Thus, when they thought about harm coming to a tree, the air, the water, the land, the animals, the birds, or to any child, they felt excruciating pain in their own bodies, just as if someone had harmed them because their cells were the cells of all other life forms. There was no longer an "us" or a "them" or even a "we". There was only "I am."

In this moment we are together writing the story of the Mayan endtime prophecy. Its conclusion is in our hands. What is certain in that story is collapse; what is yet uncertain is the extent to which collapse and global initiation will bring forth the transformation of human consciousness.

Our strongest ally in writing the new story is the earth. Or in the words of David Abrams, "To live in a storied world is to know that intelligence is not an exclusively human faculty located somewhere inside our skulls, but is rather a power of the animate Earth itself, in which we humans, along with the hawks and the thrumming frogs all participate."

Storyteller and mythologist, Michael Meade, writes in his extraordinary book, *The World Behind The World:*

> *At critical moments in history mythic sense tries to return to awareness in order to indicate life's inherent capacity for renewal. When the end seems near and nothing seems to make sense anymore, the sense of myth tries to return to make sense of all the endings and to hint at ways of beginning again. In the end there is no sense of closure, but a surprising mix of closings and openings, as the door of time swings loosely on the threshold of eternity. In the end, mythic sense returns bringing with it the wisdom of origins, the pulse of originality and the vitality of the unseen to the world-weary realm of literal reality.*[46]

I believe that if you are reading this book, something greater than your conscious mind and ego has drawn it to you. It may be that reading it and thoroughly utilizing the exercises contained within it can exponentially facilitate the transformation of consciousness that collapse is here to offer you. This book has been written with the intention of assisting you in the process of integrating your personal story with the story of the trees, the flowers, the rivers, the sky, the soil,--the light and shadow of Gaia's mysteries with those of your own soul.

You were not born in the nineteenth century or the twenty-second century; you are alive now—in this new millennium, in this place, reading this book for a purpose only your soul can know. You can deny the reality of collapse, but that will not prevent it from happening. You can continue to live the nightmare of civilization, clinging to the old paradigm, allowing the old stories to shape your life. Or, you can choose to live new stories— the story of the life/death/life cycle, alongside the story of collapse, community, and cooperation. This is the story you were meant to live—the one your soul has been whispering in your ear since you were born but which was silenced by the cacophony of empire. For all the uncertainty and angst evoked by collapse, its unsurpassable gift may be the opportunity to live and tell a plethora of stories erupting from the earth, seeding the new paradigm that begs to germinate and flourish.

REFLECTION

**What is your favorite story or fairy tale? Why do you resonate with this story?

** A time-consuming but extremely powerful and often healing exercise is the writing of one's own story. It need not be a formal autobiography, and it need not be detailed. Consider taking the time and energy to write your story over a period of days or weeks. Notice the feelings that emerge as you do this. Where in the story were you "dismembered"? Where in the story were you "re-stored" or made whole? (The word "restore" is connected with "hearing a new story", or "being re-storied", which of course, can be "restorative.")

**What was your experience of hearing the "new story" in this chapter? What would you like to add to it? In other words, what is the "new story" you would like to live?

**Take some time to reflect on the following poem by William Stafford

A Message From The Wanderer

Today outside your prison I stand
and rattle my walking stick: Prisoners, listen;
you have relatives outside. And there are
thousands of ways to escape.

Years ago I bent my skill to keep my
cell locked, had chains smuggled to me in pies,
and shouted my plans to jailers;
but always new plans occurred to me,
or the new heavy locks bent hinges off,
or some stupid jailer would forget
and leave the keys.

Inside, I dreamed of constellations—
those feeding creatures outlined by the stars,
their skeletons a darkness between jewels,
heroes that exist only where they are not.

Thus freedom always came nibbling my thought,
just as—often, in light, on the open hills—
you can pass an antelope and not know
and look back, and then—even before you see—
there is something wrong about the grass.
And then you see.

That's the way everything in the world is waiting.

Now—these few more words, and then I'm
gone: Tell everyone just to remember
their names, and remind others, later, when we
find each other. Tell the little ones
to cry and then go to sleep, curled up
where they can. And if any of us get lost,

if any of us cannot come all the way—
remember: there will come a time when
all we have said and all we have hoped
will be all right.

There will be that form in the grass.
     ~William Stafford~

NOTES

NOTES

Chapter 16–What the Creatures Can Teach Us

To regard any animal as something lesser than we are, not equal to our own vitality and adaptation as a species, is to begin a deadly descent into the dark abyss of arrogance where cruelty is nurtured in the corners of certitude. Daily acts of destruction and brutality are committed because we fail to see the dignity of Other.

~Terry Tempest Williams~

My paw is holy, herbs are everywhere
My paw, herbs are everywhere
My paw is holy, everything is holy
My paw, everything is holy

~James Koller~

A few years ago I found it necessary to euthanize a dear canine friend who had become blind and dangerously aggressive as a result of his loss of sight and the abuse he had suffered in his life before he came into mine. As he rode peacefully with me to the animal shelter where he was put down, I held his paw and tearfully recited the above chant. The memory of that day returns with deep grief in my heart every time I reflect on these words.

Every animal with whom I have developed a relationship in my adult life has been a teacher.

Each has "spoken" to me in a different way whether lying beside me as I cried, warning me of impending danger, or just looking deeply into my eyes. Similarly, animals I have encountered in the wilds of nature have instructed me, comforted me, warned me, and enchanted me. Countless studies in recent decades have revealed the uncanny intelligence and wisdom of the four-leggeds. Research with gorillas, chimpanzees, whales, dolphins, elephants, and many other species reveals their profound intelligence and intuition, as well as emotion. The *U.K. Independent* reports that "Elephants never forget the smell of a tribesman" and notes that, "Elephants mourn their dead and engage in long-distance communication using barely audible, low-frequency growls. Now they have been shown to be able to distinguish between different human tribes based on the smell and colour of their clothing." [47]

It is axiomatic that in a collapsing world, some of the worst casualties of abuse and neglect will be members of the non-human world. At this writing in 2009, we hear of the hundreds of abandoned animals left behind in the wake of massive housing foreclosures. We need only recall Katrina and other disasters where pets are abandoned or perish in heart-wrenching numbers to imagine the toll that the collapse of civilization will take on the creatures.

Michael Mountain, President and founder of Best Friends Animal Society in his cutting-edge article

"The Third Rail,"[48] emphasizes that our work for the animals is a work of the soul. He notes that whenever the news reports instances of alligators finding their way into backyard swimming pools or mountain lions attacking suburban hikers, the standard analysis is that there are "too many" alligators, or mountain lions, or whatever species may be in question. In reality, says Mountain, it's not that there are too many animals, but that there are too many humans, and "it's the overpopulation problem that no one wants to talk about." What we must do in his opinion is reduce our human population and also the population of cattle who produce vast quantities of greenhouse gasses.

During 2007 we endured the story of football hero Michael Vick who is serving a prison sentence for operating a highly lucrative dog-fighting operation. Not only is dog fighting a grotesquely bloody spectacle, but dog fight owners, including Vick, frequently hang, suffocate, or set on fire dogs who will not "perform" to their satisfaction. Such heinous, twisted acts of brutality are motivated by greed and God knows what forms of psycho-sexual gratification.

Almost daily we hear of instances where people hoard animals—dogs, cats, livestock, and do not feed or water them properly, cage them in horrific conditions, and force them to live in their own waste. In 2007, the owner of a miniature horse crammed the animal into a cage for a large dog in order to avoid airline shipping costs, and the animal was forced to endure days of ghastly confinement while traveling from Europe to the United States. The stories of animal abuse and neglect on planet earth are endless--and endlessly heart-rending.

While these stories are extreme and leave most animal-respecting individuals shuddering, we often assume that if we aren't committing these kinds of atrocities against animals, that we are guileless in abusing them. However, our consumption of factory-farmed animals is, in fact, a means of participating in civilization's abuse of animals. Sheryl Rapee-Adams, a volunteer with Best Friends Animal society in Vermont, notes that ending our consumption of animal products will enhance our bodies, our local economies, the ecosystem, and of course, animals.[49]

> *What could be healthier for humans and their local economies than eating locally-produced, plant-based foods, which also happens to prevent the torture and slaughter of factory-farmed animals and the poisoning of our arable land through genetically-engineered monoculture crops (usually soy and corn) to feed factory-farmed animals and to create highly-processed foods that cause obesity, degenerative disease, and cost us all in higher health care bills? Raising and slaughtering animals to feed humans and other animals is extremely taxing to the environment and literally takes food from*

the mouths of hungry people worldwide. It takes many times more energy to produce a pound of beef than a pound of wheat, and eating animal products is linked to numerous diseases, which raises the costs of healthcare and sometimes takes our loved ones from us. For those who eat flesh and animal products, buying those products from small farms that support local economies, pay livable wages, and maintain safe working conditions also means better lives for the animals who live and die there.

This statement honors the right of people to eat animals and animal products, but encourages us to do so in a manner that supports animals, humans, and the earth. In a collapsing world of food shortages, extremely high food prices, and contaminated food products, those who intend to eat well must grow and raise their own food organically or obtain their food from people who do, and some may also want to consider learning how to hunt game. Under such circumstances, every morsel we are fortunate to put in our mouths is likely to be consumed with great personal effort and forethought.

In recent years I have been reluctant to eat animal flesh, but since moving to rural Vermont, my perspective has shifted as I have come to understand organic farming and how differently organic farmers raise, relate to, and slaughter animals for consumption. On industrial, factory farms, animals are "mass produced" in the sense that they are birthed and warehoused in large numbers, raised on chemical and hormone-intense diets, seldom given any attention by humans other than basic custodial care, and then slaughtered mechanically with assembly line killing machines that typically involve torture or prolonged agony.

Ubiquitous stories of animal torture in the slaughtering process had left me feeling that I no longer wanted to participate in the carnage by eating beef or pork. Moreover, when I considered the deforestation that occurs in the raising of beef cattle and the hydrocarbon energy required to transport beef from a country such as Brazil to my grocery store, I found myself increasingly reluctant to consume it.

That was before I met my friend, Carl Russell, quoted above in Chapter 13—an organic farmer here in Vermont. In his 2008 article "Food From Thought" on the *Truth to Power* website, Carl shares his perspective on raising and processing animals for food and the spiritual practice it has become for him. I have included the article in its entirety below because if offers a middle ground between mindlessly consuming animal flesh produced on factory farms, or refraining entirely from eating any animal flesh.

When I was a teenager I participated in a group of friends who were enthusiastic about our outdoor adventures. Fishing, hunting, hiking, and camping, we immersed ourselves in natural experience. Like many of our kind, during summer we would seek deep cool water to recharge our spirits. One favorite swimming hole was in an abandoned copper mine on the side of a mountain, several miles from town. A jeep trail led there through challenging terrain, enhancing the adventure.

During mining operations copper ore had been blasted out of the bedrock, leaving long narrow ravine-like shafts. Once abandoned they had become filled with water. The steep rock ledges were burnt-orange, almost red in color. The water was bright aqua-blue, and milky with suspended sediment. The contrast between green forest, red earth, and brilliant blue water created an exotic visual effect.

Upon arrival we would race over the barren ground to the edge of the cliffs, and plunge one behind the other into the cold blue water. Once we calmed down from the initial rush, we would engage in the main purpose for our coming, cliff jumping. There was an increasing gradient along one side of the mine where we could jump from spots ranging in height from ten, to as high as sixty feet. We would freely charge out into the air from the lower cliffs, demonstrating different styles of cannon balls and dives. The approach at the highest place was more subdued. The cliff walls of the ravine were only thirty feet apart, and from a height of sixty feet an aggressive jump could end dangerously close to the opposite side.

I never found it easy to jump from the high cliff. I knew that I didn't have to make the jump, but something inside me encouraged me to try. I would take my time ascending, and once on the rock platform, I would adjust to the challenge in front of me. From above, the chasm seemed deceptively narrow, and as I looked down I would lose my depth perception. I could not see beyond the surface of the milky-blue water, so the view took on a two-dimensional appearance. Light reflecting from ripples would shimmer hypnotically, making the water level seem to rise and fall, like ocean swells.

Finally, I would be compelled to step off into thin air. The step was my last conscious act. The decent was so rapid that there was no time to think. I was completely dependent upon my instincts to keep upright and prepared for submersion. I can still remember the sound of the air ripping past my ears, and the sensation of my body tearing through the water. Gradually slowing down, then regaining buoyancy, I had a sense of exhilaration as my mind caught up with my body, mentally absorbing the experience.

I found myself recalling these memories as I stood outside the pen where I raised my first pigs for slaughter. I had decided to move ahead with a challenge that had been rising within me for years. I felt the need to raise and slaughter animals for my own meat consumption. My parents had insisted that as a young hunter I eat everything I killed. This helped in part to shape the current motivation, but it had also tempered my desire to kill things. A bird on the wing, or a white-tail deer at fifty yards, is quite different than a pig at hand, and I needed to adjust to the challenge in front of me.

As I readied myself to enter the pigs' pen, I could see the shimmering in their eyes, too close, too far, too narrow. I was standing at the cliff's-edge of a set of experiences, the depth and breadth of which I could not fathom. Even though I knew that I didn't need to make the choice to kill my pigs, I was compelled to. To this day, I have no idea why I trusted myself to take that step, but I entered the pen and did what needed to be done.

I had started this endeavor because I firmly believed that if I was going to enjoy animal flesh for food, then I had to take responsibility for the killing. The act of killing became complicated when I realized how important the pigs' lives were to me. These two distinctly different feelings were difficult for me to wrap my mind around. I faced an intellectual chasm, and I could sense the uncertainty that eddied there. After twenty years I am still in awe of the world that opened up around me as I tumbled over that precipice.

Killing animals requires skill and commitment. I started out with some skill, but more commitment. There are many aspects of slaughtering and butchering that can only be learned through experience, and after several years slaughtering chickens, cows, and pigs, I started to feel comfortable with the process. I felt myself regaining buoyancy, and I began to realize how fast my life had been changing. I had been acting more from instinct than from conscious thought, and I had become submerged in experiences that my mind was only beginning to absorb.

There are many emotional issues surrounding the care, and consumption of animals. Because they move, and breathe, and make noise, we can relate to all animals on a most basic level. Whether cat, or deer, chipmunk, draft horse, or milk-cow, we can empathize with their life experience. It is enjoyable to husband farm animals because we can create relationships with them, which enhance our own emotional lives. The recognition of the value of these relationships to my life is what compelled me to start raising a diversity of animals on my small farm.

Beyond feeding, and cleaning pens, I involve myself in the lives of my animals. I spend time with them, scratching their itches, encouraging them to play, looking into their eyes, and touching them compassionately with friendship. They respond positively, becoming enjoyable engaging creatures. When they are alive, it is important to me that they are relaxed and comfortable, and that I have taken time to have relationships with these living being.

I know many people who would rather not know the animals whose meat they may eat. It is common for them to acknowledge that they would become too emotionally attached. Often the result is that the meat they do eat comes from animals raised in confinement and ignored by humans until their meat is processed. This choice is right for a lot of families because it is convenient, and many people don't have the space to raise animals, but I am concerned about the loss of the value of the relationship.

To me food is more than calories or culinary preference. Food is the energy of life, and life is about relationships. Life is the flow of energy from the Earth through all things. It is the relationships between all things that keep the energy flowing. Like everything else, humans gather energy, then expend it, and eventually we return in totality to the source. Along the way, it is the relationships that we make that define our participation.

Once I recognized the significance of my need to have relationships with the animals whose flesh I eat, I began to truly feel my connection to Life energy. I feel the same relationship to the trees in the forest where I work, to the soil that I cultivate, and to the plants in my garden. It is clear to me now that energy does not leave an animal when it dies, nor does it leave a bean picked from a vine. The energy is always there. By involving myself physically and emotionally in the production of my food, I can strengthen my relationship to that energy. I also believe that my involvement can help these animals to have vital lives, enhancing the value that the food, in turn, brings to my life.

The food that we raise on our small farm recharges my body, and my spirit. I am proud of the loving relationships we have with our animals. Being the person responsible for taking their lives is part of demonstrating my commitment to them. Acknowledging emotional investment substantiates the value that I place on their existence. The relationships to my family, to our community, and to the Earth are strengthened every day through this connection to food. These animals and plants that we raise fuel our efforts to deepen our involvement in the energy of Life from one season to the next.

While I write this, I am overwhelmed by emotion, and I feel exhilarated as I float in the realization of personal experiences, absorbing the meaning. There are many people who, for one reason or another, cannot engage in these experiences, but I encourage those who think that they can, to try. It is the fear of uncertainty that causes discomfort, not uncertainty itself. I have found that I can be afraid, or I can trust myself, but uncertainty will always exist. Uncertainty is what fills the gap where relationships grow. By diving into that abyss I have found a pool of significant reward, and I feel recharged as I watch the ripples form from my wake and spread across the life around me.

Respect, A Poem By Carl Russell

Conceived, born, and raised on this land,
the farm of your dam.
Where you tested your feet,
and found your first teat.
Many years had passed,
since they'd seen the last,
Jersey bull hazing,
this hillside grazing.
From you period of toil,
the wind, water and soil,
of Gilead are within you.
We are not taking this life,
or energy from you,
it will always be yours.
We merely use it now,
to feed our bodies,
and to fuel or minds,
to manifest our dreams,
and to empower our values,
to perpetuate the care of this land,
and your kin who will follow.
Through you we touch this soil,
to the very heart and spirit of the Earth.
You help us to be part of the system,
allowing us to become products of our own work.
The Earth as my canvas,
Life is the easel,
My being the brush,
I paint my dreams.

A number of other organic farmer friends here in Vermont have enlightened me regarding the importance of understanding where my food comes from by knowing who raised it and how. The essence of the relocalization movement that is sweeping the nation is that place and people are fundamental in the production and consumption of food. This is the antithesis of empire which commodifies animals by turning them into products to be mass produced, transported to distant regions of the world, and consumed as we might any other item in a capitalist system.

Our indigenous ancestors knew well the capriciousness of nature and gave profound thanks through ritual and prayer to the animals and plants they killed or grew for food. They lived according to the notion expressed by David Stendl-Rast that nothing can be "taken for granted", especially in times of famine, for in those conditions, every form of sustenance one is able to obtain is "granted", that is, a gift.

We should not be surprised at the hideous maltreatment of animals in the culture of empire. After all, civilization is about objectification and ownership. People, animals, and the land are "things" to be dominated, consumed, or profited from, not fellow creatures with whom to develop a relationship. While women, children, and the elderly are considered expendable in civilization, animals are deemed even more disposable.

On the other hand, I have never met anyone who is actively preparing for collapse who is not fond of animals. Perhaps that is because those individuals have allowed the non-humans to teach them about vulnerability, loyalty, sacrifice, risk, danger, loss, affection, and much more. Perhaps it is because something in us knows that 200 of them go extinct each day, and we are yet another species, unable to predict when the collapse of civilization will manifest *our* extinction.

Because negativity is unknown to them, the creatures are superb teachers, says Eckhart Tolle, commenting that:

> *No other life form on the planet knows negativity, only humans, just as no other life-form violates and poisons the earth that sustains it. Have you ever seen an unhappy flower or a stressed oak tree? Have you ever come across a depressed dolphin, a frog that has a problem with self-esteem, a cat that cannot relax, or a bird that carries hatred and resentment? The only animals that may occasionally experience something akin to negativity or show signs of neurotic behavior are those that live in close contact with humans and so link into the human mind and its insanity.*

> *Watch any plant or animal and let it teach you acceptance of what is, surrender to the Now.*

> *I have lived with several Zen masters—all of them cats.*

We will become more teachable and develop more capacity to learn from the creatures as we ourselves become more "creaturely." Civilization has disowned wildness as primitive and inimical to "progress." It has inculcated shame in us regarding our bodily functions, fluids, odors, and needs. Driven by religion and the Puritan ethic, our sexuality is often estranged from our innately animal instincts. To be "civilized" is synonymous with being domesticated, restrained, and repressed, and if we participate in sexual behavior at all, we are encouraged to do so in a controlled, sanitized, or even surreptitious fashion.

Obviously, functional human beings interacting with other human beings must adhere to limits and boundaries as emphasized in Chapter 5; paradoxically however, the more we celebrate our inherent animal nature, the more likely we are to effortlessly honor our limits. We have only to notice that it is not animals that are soiling their nests and desecrating their habitat, but rather humans. As Tim Bennett states in "What A Way To Go", the ant population of our planet is astronomical, but ants are not building 6,000 square foot houses or buying plasma TV's. Animals live within the limits of their environment because their lives depend on doing so.

In Chapter 13, I spoke of my interview with Lisa McCrory and Carl Russell in relation to raising children in a collapsing world. My original interest in Lisa and Carl, however, was evoked by their organization Earthwise Farm and Forest in Randolph, Vermont, where they use only draft animals in their organic farming operation. Their intention in using horses and oxen rather than tractors and other fossil fuel-powered machinery is not only to conserve energy, but being well-aware of collapse, they realize that at some point, gasoline will become unaffordable, and ultimately, in a post-collapse world, farm machinery itself will not be available. Carl, Lisa, and other organic farmers prefer to use draft animals for plowing and hauling because of essential tasks which can be performed more efficiently by animals than tractors. When doing so, however, it becomes essential to develop the kind of relationship with draft animals described above by Carl.

A post-petroleum world is likely to be one where mobility will be strictly limited to walking, horseback riding, or riding in animal-drawn carts with wooden wheels. Large-scale farming will not be possible unless draft animals are used. Consequently, humans will become dependent on animals for transportation and energy as we have been in earlier times, and this will undoubtedly transform our relationships with them.

Collapse will offer more opportunities than we might imagine for connectedness with other species. From house pets to draft animals to those we must kill and eat to those that will tell us what time of day or what time of life it is, the creatures must and will be our companions. They have lessons to teach us. Can we listen and learn?

Civilization has given us "dominion" over the earth and its creatures, but in the new paradigm the creatures will be the elders, the truly wise ones who have come to help us remember our animal origins and our animal destiny as the poet W.S. Merwin so eloquently writes:

> *Elders,*
> *We have been here so short a time*
> *And we pretend that we have invented memory*
> *We have forgotten what it is like to be you*
> *Who do not remember us*
> *We remember imagining that what survived us*
> *Would be like us*
> *And would remember the world as it appears to us*
> *But it will be your eyes that will fill with light*
> *We kill you again and again*
> *And we turn into you*
> *Eating the forests*
> *Eating the earth and the water*
> *And dying of them*
> *Departing from ourselves*
> *Leaving you the morning in its antiquity.*

REFLECTION

**What animals, wild or domestic, have had the most influence in your life?

**What kinds of lessons have you learned from animals?

**Take some time to journal about your animal self—your animal body, your animal emotions. What about your animal nature causes you to feel ashamed? What about it causes you to feel joy, courage, freedom, peace?

**What do you know about your own "wild" nature? What frightens you about it? What do you love about it? How has your wild nature been useful to you in the past? How could it be useful to you in the future?

**Have you ever killed an animal? If so, journal about the experience. If you have not, what do you feel when you think about killing an animal for food?

NOTES

NOTES

Chapter 17–In Sickness and in Health

We might imagine much of our current disease as the body asserting itself in a context of cultural numbing. The stomach takes no pleasure in frozen and powdered foods. The back of the neck complains about polyester. The feet die of boredom for lack of walking in interesting places. The brain is depressed to find itself described as a computer and the heart surely doesn't enjoy being treated as a pump....We are perhaps the only culture to regard the body with such poverty of imagination.

~Thomas Moore, *Care Of The Soul*~

As I write these words, CNN is featuring its two-part series "Planet In Peril" which is tracking in depth the global devastation now intensifying as a result of climate change and energy depletion. Little in the series is new to me, but one fact has leapt out at me, namely, the reality that as millions of plant species are destroyed, so will be hundreds of sources for producing medicinal remedies. With that, the reality and brutality of collapse just hit me over the head one more time.

For years I've known that as civilization crumbles and the earth is increasingly ravaged by the repercussions of it, commercial medicines will become extremely scarce or completely unavailable. Yet I've cherished the hope that medicinal plants could be harvested by individuals to treat illnesses that vanishing medications once alleviated. After all, isn't there a certain romance or sense of empowerment in being able to grow one's own herbs and treat one's own ailments or those of one's loved ones? Perhaps it's the ancient "witch" in me who revels in stories of crones mixing potions and elixirs for the community. But that was then; this is collapse. You see, the herbal species are going extinct as well.

One momentous casualty of collapse will almost certainly be health care. It is now fading from accessibility for most working and middle class Americans, and soon it will be only a luxury of the very wealthy. Above I have mentioned the necessity of learning first aid and alternative healing skills such as herbal medicine, Reiki, energy balancing, massage, homeopathy, acupuncture, and others because as the health care system crumbles, the world will be in dire need of healers.

A treasure-trove of information is available at Peak Oil Medicine[50], a website managed by Dr. Paul Roth which explains not only the realities of healthcare post-collapse, but offers practical suggestions for families and healthcare professionals in terms of preparation. In an October, 2006 article[51] by Roth, he notes the usefulness of applying permaculture principles to the concept of healthcare after Peak Oil. (For an in-depth understanding of permaculture, see the website of the Permaculture Institute.[52]) Following is a portion of Roth's healthcare paradigm, largely informed by the fundamentals of permaculture:

Permaculture-inspired ideas for healthcare after peak oil

We will need to look at the big picture first, and not get lost in the details of a solution. The strategies used at each location will be different, and will likely need to be adapted to changes that occur over time (for example if there is a sudden influenza epidemic, severe drought, or other catastrophe).

As permaculture uses "self-maintaining systems", the implication is that each individual will need to take more responsibility for their own body, and try to be as healthy as possible. There will need to be a change in focus from the treatment of disease to the promotion of wellness. This idea is derived from the principle of minimizing waste, as it is wasteful to use scarce healthcare resources treating a preventable disease.

The system will also need to allow for changes in illness patterns. On the one hand, people are likely to be much more active, eat less processed food and lose weight. On the other hand, accidents, musculoskeletal injuries and infectious diseases may be more prevalent.

Additionally, it will be important to enlist the whole community in achieving good health, and the current boundaries that separate medical workers from the general public will become blurred.

Sustainable healthcare systems will probably include plant-based treatments (based on the ability of plants to catch and store solar energy). Holmgren says that "herbal medicine might not provide a complete pharmacopoeia, but we can, to a very great extent, successfully treat many ailments with locally grown and processed botanical medicines." While you may or may not agree with this assertion, it is the idea behind it that is important: that locally produced things can fix health problems.

The focus on diversity and small-scale and slow (or lower-tech) solutions is based on Schumacher's work. It is a concept that supports relocalization, and the judicious use of technology on an appropriate scale (perhaps using a microscope to check a urine specimen for infection in a doctor's office, rather than sending the specimen off to the lab for culture).

A negative implication of diversity is that solutions will need to be designed to resolve a variety of problems unique to each location. An example: Distribution patterns of mosquito-borne illnesses like dengue fever and malaria are likely to alter as climate change accelerates, possibly making them a major problem in one location but not another. The diversity principle also suggests that medical systems will need to be designed with built-in flexibility to handle emergencies and other unforeseen events.

Appropriate technology

Schumacher discussed his ideas in his book *Small is Beautiful: A study of economics as if people mattered.* He believed that "production from local resources for local needs is the most rational way of economic life." Appropriate technology uses the minimum level of complexity required for the job at hand. It ideally can be made locally (or at the least maintained and repaired there), is of low cost and requires little maintenance.

For our discussion, appropriate technology should be made from locally available, sustainable materials, and contain little or no oil derivatives. There are many examples of this technology related to healthcare; the main ones are *in public health areas like sanitation and clean water provision.*

A final idea of Schumacher's is that the reduced efficiency arising from using appropriate technology necessitates more human labor to produce a given amount of goods. This ensures full employment (thereby occupying otherwise idle workers) and is theorized to promote health, beauty and permanence.

Summary

Following oil peaking, we can choose to allow our society to slide into anarchy (as has Zimbabwe, and to a lesser extent Russia). Or we can choose an ethically-based and eco-centric pathway leading to a compassionate, humane and richer society typified by clusters of small-scale, self-sufficient communities. The choice is ours.

Healthcare is one of the more formidable aspects of navigating collapse, and one that many people cannot bear to consider. Overall, the lifestyle of individuals living close to the earth, eating organic foods, doing robust physical labor because their survival depends on it, and not being subjected to the lifestyle of civilization will promote wholeness and serve to prevent illness. However, families and communities should have large quantities of medical supplies on hand, as many natural remedies as possible, and as many traditional medicines as they can access.

Attending to the illnesses of others will be a necessity in any collapse-conscious community. Ideally, these duties will be shared and emotionally processed in regular dialog circles. As with any adversity, illness is present not only to offer soul development to the person who is ill but to everyone around him. Practically speaking, communities that survive collapse are not likely to have an actual physician in their midst. They may be fortunate to have a registered nurse or an individual in the community who has had some medical or dental training. Nor are they likely to have access to medical equipment or supplies and will probably have to remove them from abandoned hospitals or clinics.

In 2007 we began to witness the proliferation of global pandemics and viral and bacterial infections resistant to treatment of any kind. Outbreaks of staphylococcus infections nationwide were resulting in schools being evacuated and closed down for days and weeks. My colleagues in the institutions where I regularly taught became obsessed with hand-washing and using anti-bacterial wipes and sprays to protect themselves from contamination. Clearly, the immune system of the planet is screaming its distress signals, overwhelmed and toxic as it has become as a result of civilization's befouling behavior. As the immune system of the planet deteriorates, so do the immune systems of humans and animals, making all of us far more susceptible to virulent, incurable diseases.

As I have reiterated throughout this book, the outcome of collapse is appallingly uncertain, and the survival of humanity, as well as all other species which have not yet gone extinct, is at best, precarious. The most meticulous preparation for collapse cannot guarantee longevity, let alone the likelihood of a flourishing, fulfilling existence. None of us knows who of us will or will not survive. As noted above, we are indeed like passengers on the Titanic with an inadequate supply of lifeboats. As we build our own, we do everything humanly possible to ensure our well being, yet knowing that our best efforts may be inadequate.

Probably the most fundamental reason that the majority of the human race is unwilling to come to terms with collapse is that collapse is indeed a form of death which forces us to confront our own mortality. No sooner do we begin discussing collapse than we are staring into the black maw of our final days on earth, even if it turns out that we will live another half-century. The last chapter of this book will address the issue of death— our own and that of our loved ones, but in this chapter I do not assume that sickness and death are synonymous. Anyone who ventures into the wilderness may get sick or injured, but common sense and mindful preparation can often alleviate life-threatening situations.

In addition to the suggestions offered above regarding learning alternative healing techniques, I strongly recommend four publications by the Hesperian Foundation[53]: *Where There Is No Doctor, Where There Is No Dentist, Where Women Have No Doctor,* and *A Book For Midwives*--books written for lay persons living in areas where healthcare is not accessible. While communities during and post-collapse may be able to organize their own clinics, and doctors and nurses may be able to do something they haven't been able to do for decades, namely, make house calls, it behooves every individual and household to learn healthcare skills and have access to "how to" materials for treating injuries and illnesses.

Equally urgent is preparation and prevention through diet and exercise. The latter in a post-collapse world cannot be based on daily visits to the gym or reliance on sophisticated exercise machines. Those who are not involved in taxing manual labor must attend to exercising regularly which probably will mean a daily regimen of walking or swimming if clean rivers or streams are available. A daily or twice-daily meditation practice for relieving stress is also, in my opinion, not a luxury but a necessity. I have maintained my own personal meditation practice for 30 years and attribute to it much of my stamina and ability to weather various forms of stress both past and present.

In families or living communities we may be faced with long-term care of the sick and aging. Our living communities will undoubtedly be our own hospitals and nursing homes which will necessitate round-the-clock care in many instances. Individuals will be required to spend extended periods of time with friends who are ill, infirm, or dying. The emotional toll this will take on households and communities will be staggering. For this reason, the kind of circle work described in Chapter 12 will be absolutely essential to emotionally support caregivers and everyone touched by their efforts. Circles may need to meet several times a week or even daily in order to process feelings and discuss logistics.

While everything I have just described regarding healthcare during or post-collapse may sound overwhelming, and so it may be, I can also foresee opportunities for deep connection, solidarity, compassion, mutual support, and the discovery of deeper layers of "caring for the sick" than any of us has ever known. Within the bleakest scenarios may lie enormous potential for profound healing of everyone sharing in caregiving or supporting caregivers, as well as for the patient.

In *Care Of The Soul* Thomas Moore suggests in his chapter "The Body's Poetics of Illness" that we need to look into the mythology of our illnesses and consider them from the soul's point of view. From that perspective, says Moore, "Our wounds remind us of the gods. If we allow sickness to lead us into wonder about the very base of experience, then our spirituality is strengthened....Illness offers us a path into the kind of religion that arises directly from participation in the deepest levels of fate and existence."

From Moore's perspective which echoes both Buddhism and Jungian psychology, suffering and illness are inherent in the human condition; therefore, the illness serves a purpose by coming to us as a teacher. If we try to understand the cause of the disease or obsessively attempt to prevent illness, we may miss the message that the illness has come to communicate to us. Within the illness is an opportunity for a more intimate connection with the sacred or as Moore states:

The point is not to understand the cause of the disease and then solve the problem, but to get close enough to the disease to restore the particular religious [spiritual] connection with life at which it hints. We need to feel the teeth of the god within the illness in order to be cured by the disease. In a very real sense, we do not cure diseases, they cure us, by restoring our religious participation in life.

In other words, Moore is suggesting that because we are "cured by our diseases" our obligation is as much to be taught by them as it is to eradicate them. If this notion is valid, then the illness brings a message of "cure" not only to the patient, but to the greater community. As Moore notes, the word "cure" has its origins in the word "care". In ancient times curing was inextricably connected with caring, and in fact, curing did not so much mean eradicating as it meant attending. Thus, in a post-collapse household or living community, all members must be open and attentive to the disease's "cure" as they care for the ailing individual. Their commitment must be not only to providing care but to humbly asking the question, probably many times, "What has the illness come to reveal to us?"

In the heroic culture of empire it is extremely difficult to accept limits of any kind, but especially the limits of mortality. Yet given the uncertainties inherent within collapse, we are likely to share space with people we do not know well or know at all. We may share in their living and also in their process of dying as they may share in ours. Illness, incapacitation, and disability are some of the lesser deaths mentioned above that prepare us for the ultimate death. While it is important to be familiar with remedies and procedures that may eradicate illness, it is equally important to acquire the capacity to surrender to death when we have exhausted all attempts to sustain life because it may be the ultimate and most merciful gift we can give to the ones for whom we are caring, and to ourselves.

REFLECTION

**What am I doing in present time to attend to my personal health and physical well being in preparation for collapse? If certain medications are necessary for my continued health and well being, what am I doing about acquiring large quantities of those or experimenting with natural remedies that may replace the use of medications?

**What limitations do I experience in my mobility and physical activity? Is there anything I can do about those? If so, what do I need to do?

**What illnesses or physical constraints am I struggling with? How will I attend to those in a post-collapse world? Are there medical or dental procedures that I need to take care of now before medical and dental care become inaccessible?

**It may be helpful to spend some time sitting quietly with eyes closed, tuning in to parts of the body that are in pain or where illness or limitation reside. In that quiet space it is very important to listen to one area of the body at a time that may be "speaking" through illness. As one listens, it is useful to notice images that appear as well as emotions that surface. Equally important is journaling about or drawing the images.

**What has been my experience in caring for sick people? What has been draining? What has been energizing? What skills do I bring to the care of the sick?

**What healing skills have I learned? What healing skills am I drawn to learn? Have I begun collecting medical, first aid, or alternative healing supplies?

EXERCISE FOR BECOMING PRESENT IN THE BODY

Eckhart Tolle in *The Power of Now* explains the crucial importance of consciously inhabiting the body. Whereas some spiritual teachings, including traditional Christianity, have encouraged devotees to abdicate the body and aspire to dwelling only in the spiritual realm, Tolle, in the Buddhist tradition, admonishes us to become deeply present in the body as a means of fully experiencing the power of the present moment. While it is true that we are more than our bodies, it is also paradoxically true that the most efficient means of discovering that we are more than our bodies is by becoming thoroughly present *in* them.

The mind and human ego, which often become the cruel instruments of empire, tend to disengage our attention from the body, preventing us from being present in it. What is more, deep presence in our bodies is as profoundly effective in strengthening the human immune system and increasing longevity as it is in assisting us in experiencing advanced states of consciousness. Eckhart Tolle suggests, in fact, that, "The key is to be in a state of permanent connectedness with your inner body—to feel it at all times." He explains that the inner body is simply the energy field inside the body which, when consciously felt, allows us to experience our connectedness with everyone and everything else in the universe. A full explanation of the inner body and the importance of being present in it is given in *The Power of Now*. Following is one exercise offered in the book for enhancing presence in the body:

> *When you are unoccupied for a few minutes, and especially last thing at night before falling asleep and first thing in the morning before getting up, "flood" your body with consciousness. Close your eyes. Lie flat on your back. Choose different parts of your body to focus your attention on briefly at first: hands, feet, arms, legs, abdomen, chest, head, and so on. Feel the life energy inside those parts as intensely as you can. Stay with each part for fifteen seconds or so. Then let your attention run through the body like a wave a few times, from feet to head and back again. This need only take a few minutes or so. After that, feel the inner body in its totality, as a single field of energy. Hold that feeling for a few minutes. Be intensely present during that time, present in every cell of your body. Don't be concerned if the mind occasionally succeeds in drawing your attention out of the body and you lose yourself in some thought. As soon as you notice that this has happened, just turn your attention to the inner body.*

Tolle also suggests that whenever one finds oneself in a situation where one is forced to wait, such as in a check-out line or while sitting in traffic, one should seize the opportunity to practice presence with the body. In those moments, one can breathe deeply and consciously, feeling the abdomen expanding and contracting, feeling the air passing in and out of the lungs. Then feel the inner body. In this way, says Tolle, "...traffic jams and lines become very enjoyable. Instead of mentally projecting yourself away from the Now, go more deeply into the Now by going more deeply into the body."

NOTES

Notes

CHAPTER 18–SOULSICKNESS AND EMOTIONAL COLLAPSE

I am not a mechanism, an assembly of various sections.
And it is not because the mechanism is working wrongly
that I am ill.
I am ill because of wounds to the soul, to the deep
emotional self
And the wounds to the soul take a long, long time, only
time can help
And patience, and a certain difficult repentance,
Long, difficult repentance, realization of life's mistake,
and the freeing
Oneself from the endless repetition of the mistake
Which mankind at large has chosen to sanctify.

~D.H. Lawrence~

A mosaic is a conversation between what is broken

~Terry Tempest Williams~

Although I am currently an adjunct professor of history, an author, and enjoy managing my own website, seventeen years of my life were devoted to the psychotherapy profession where I sat daily with countless individuals suffering from wounds to the soul. My clients taught me more than I could have ever taught them, and I cherish the years in which I walked with them through their emotional pain. Most were courageous individuals who were willing to tell the truth about their wounds, their families of origin, and the society in which they lived. For this reason, I imagine that they will be better equipped emotionally to navigate collapse than the majority of Americans who have little if any conscious awareness of the demise of civilization and the consequences it will entail for them.

As I interact with others who are well aware of collapse and even preparing for it, I am most frustrated by those who focus on the logistics of preparedness in terms of learning skills, relocation, food storage, and other aspects of navigating collapse, yet ignore or neglect the emotional and spiritual dimensions of it or insist that those are extraneous luxuries with which we cannot afford to be concerned. Some of these individuals still believe that they can survive in isolation and recoil when the topic turns to community and interdependence. And of course, my atheist and agnostic friends predictably roll their eyes when I mention "sacred demise" while others tell me that in a post-collapse world, people will find their own spiritual path.

Admittedly, it is far easier to pre-occupy ourselves with logistics than to ponder the emotional and spiritual realities of collapse. What passenger on the Titanic when it was time to count the lifeboats would have stopped, breathed deeply, and asked herself how she was feeling in the moment? It is always much easier to "manage" a crisis than to be present with it. Yet, the agonizing truth is that while many individuals will be able to physically navigate collapse, some will not be able to do so emotionally. There will be countless emotional breakdowns and suicides as the wheels fly off of empire. In my opinion, collapse will become psychologically intolerable for those who have no inkling of it, who are emotionally tethered to possessions, status, careers, and lifestyles that provide identity and security. Nevertheless, many individuals for whom collapse will be no surprise and who believe they are prepared for it may also succumb to psychological meltdown—especially if they have spent a great deal of time rolling their eyes when the spiritual and emotional aspects of collapse are mentioned.

How then do we cultivate emotional and soul health in present time which will serve us in more dire segments of collapse?

Some people have found individual psychotherapy useful, especially when the therapist's perspective encompasses social and cultural issues as well as the client's personal history. Sadly, the majority of therapists are not up to speed on the realities of collapse, and as a result, may pathologize a client who is and who freely speaks their perceptions about collapse in therapy.

I frequently hear stories from individuals who have been working with a therapist for some time, and after learning more about Peak Oil, climate change, and other topics related to collapse, feel compelled to share their concerns with a therapist who is not informed on these issues. While the therapist may be remarkably kind, compassionate, and well-trained in her field, she or he may not be educated in the realities of collapse. In fact, it is not uncommon for some therapists to arrive at their office in their SUVs decorated with bumper stickers advertising their favorite progressive candidate yet listen with skepticism to a client discussing his concerns about collapse, only to unconsciously tune that person out and attribute the client's concerns to his or her "issues." I particularly recall one such horror story in which a female friend of mine shared with her therapist her concerns about Peak Oil and was told that she would benefit from a prescription for the anti-anxiety medication, Xanax.

When I hear such stories, I am reminded of the remarkable 1992 book by authors James Hillman and Michael Ventura *We've Had A Hundred Years of Psychotherapy and Everything Is Getting Worse* in which the authors agree that psychotherapy in its present form is inadequate to address modern anxieties and neuroses. However, a few clinicians practicing psychotherapy *are* awake and as a result, are able to be present with clients who are becoming increasingly aware of world events and are experiencing what has become known as eco-anxiety.

One notable example is an article entitled "The Waking Up Syndrome", written by the author of this book's foreword, Sarah Edwards and her colleague, Linda Buzell-Saltzman, who are wise and well-informed eco-psychologists, consciously preparing for collapse. Sarah and Linda are not only actively practicing ecopsychology and informing the world regarding collapse issues, but because of their own capacity to examine collapse are able to be empathically present with clients who need to share their feelings and thoughts about it. In a December, 2008 blog post, Sarah Edwards traces the evolution of the term "eco-anxiety" and notes that "...the term eco-anxiety has become integrated into normal parlance, taken more or less as a given of our time. You'll notice also that rather than its being cast in the pejorative, more often than not it is being used in conjunction with tips and ideas for what someone can do about concerns one feels about such things as peak oil, climate change, and environmental degradation."[54]

Yet while some therapists such as these women are tuned in to collapse, psychotherapy itself has become as inaccessible in the past two decades as quality medical care in the United States. Nevertheless, if one has access to a well-informed therapist and the means with which to afford the work, it can offer invaluable emotional support and healing.

Ongoing circle work as explained in Chapter 12 is inexpensive and provides group support and the opportunity to process feelings with others who share a similar perspective and similar concerns. In my opinion, the sooner one can find and participate in a circle, the better. We should not wait for the more severe aspects of collapse to erupt before garnering this kind of support, and as the intensity of collapse increases, so will the frequency with which circles need to convene. A list of resources at the end of this book includes information regarding finding or creating such circles.

In addition, we must begin now to utilize other tools such as art, music, drumming, story, poetry, and ritual to support our own psyches and those of our community as we all experience personal and global initiations. As with circle work, all forms of support must be increased and deepened alongside the accelerating losses of humanity and the earth in all stages of collapse.

Ritual is exceedingly important in stabilizing the emotions. In Chapter 10, I spoke of the healing power of the grief ritual, but it is only one of countless rituals that individuals and groups may utilize. I mention ritual above in the context of various forms of art because the word *ritual* has its origin in an Indo-European root meaning "to fit together." Anthropologist and educator, Angeles Arrien states in her book *The Fourfold Way* that ritual is related to art, skill, order, weaving, and arithmetic—"all of which involve fitting things together to create order." Furthermore, says Arrien, "Ritual is the conscious act of recognizing a life change, and doing something to honor and support the change through the presence of such elements as witnesses, gift giving, ceremony, and sacred intention."[55]

Ritual, in order to meaningfully support us, must be spontaneous, and in the words of Malidoma Somé, must "arise out of the earth and the psyche" as opposed to being contrived or adopted from written instructions. Only in this way can ritual assist us in "fitting together" the parts of ourselves that may be crumbling as the world around us crumbles.

Yet for all the emotional preparations we may employ, collapse and its resultant losses will be psychologically ruthless for most individuals. Unquestionably, emotional breakdowns will become commonplace, if not within our households and living communities, in the larger community around us. Depression will be epidemic, and few of us are likely to escape it. While I am not averse to the moderate, carefully-monitored use of antidepressants, they will be very difficult to obtain during and post-collapse, and therefore, it behooves us to be familiar with herbal and other natural remedies for treating depression. Even more salutary will be the proximity of loving, nurturing, inspiring, and stimulating support in the form of family or friends.

Both James Howard Kunstler and Dmitry Orlov in their writings about collapse paint a grim picture regarding the ability of individuals to cope with the trauma of collapse. Kunstler conjectures by way of his fictional novel that madness and suicide will be commonplace in a collapsing world, and Orlov's comments based on historical record regarding collapse in the Soviet Union appear to substantiate Kunstler's projection.

According to Orlov, the early stages of collapse are likely to be the most chaotic, and he recommends that if aware individuals can find a place to temporarily withdraw from it, they are likely to be less stressed; however, at some point, they will find it necessary or may simply choose to return to the milieu after the situation has stabilized. But while escaping the throes of chaos may alleviate a great deal of stress, returning to a collapsed world may itself be traumatic as one mentally and emotionally absorbs the losses that have occurred.

Just as we are likely to find ourselves caring for those who are suffering physically as collapse unfolds, we are also likely to find ourselves attending to friends and loved ones who have been emotionally devastated and need our nurturing support. Collapse will take an enormous toll on our souls as well as our bodies—yet another reason why emotional and spiritual preparation dare not be minimized.

REFLECTION

**What am I doing to prepare emotionally for collapse? What kind of support systems do I have in place for the transition?

**What would be the greatest losses for me during and after collapse? How would I take care of myself if I were to experience such losses?

**What tools such as drumming, story, poetry, art, or music am I using to support myself emotionally in the current moment and how might these tools assist me as collapse intensifies?

**What rituals have I found helpful in sustaining me during loss in my life?

**Do I have people in my life with whom I can talk openly and intimately about collapse? If so, what is it like to have such relationships? If not, what do I need to do to find and develop that kind of connection?

**Yes, dear reader, I'm asking this question again: Do you have a spiritual path, a connection with something greater than yourself that adds meaning and purpose to your life? Spend some time journaling about or drawing this connection. (If you do not have such a connection, journal about or draw the absence of it. What would it take to develop a relationship with something greater?)

NOTES

Notes

Chapter 19–Hospice As Holy Ground

When death comes
like the hungry bear in autumn;
when death comes and takes all the bright coins from
    his purse
to buy me, and snaps the purse shut;
when death comes
like the measles-pox;
when death comes
like an iceberg between the shoulder blades,
I want to step through the door full of curiosity,
    wondering:
what is it going to be like, that cottage of darkness?
And therefore I look upon everything
as a brotherhood and a sisterhood,
and I look upon time as no more than an idea,
and I consider eternity as another possibility,
and I think of each life as a flower, as common
as a field daisy, and as singular,
and each name a comfortable music in the mouth
tending as all music does, toward silence,
and each body a lion of courage, and something
precious to the earth.
When it's over, I want to say: all my life
I was a bride married to amazement.
I was the bridegroom, taking the world into my arms.
When it is over, I don't want to wonder
if I have made of my life something particular, and real.
I don't want to find myself sighing and frightened,
or full of argument.
I don't want to end up simply having visited this world.

~Mary Oliver~

I occasionally receive hate email but more frequently receive ones like this: "I've just unsubscribed to your email list. Your website is filled with negative stories and articles, and I need to keep a positive attitude and do what I can to make my world better."

How does one describe the tone of such a statement? Angry? Not really. Disappointed? Perhaps. Scared? Probably. But I think that *righteous* is the word I would use to describe this reader's perspective. By righteous, I mean a false sense of doing or feeling "the right thing", but the problem with a righteous attitude is that it often leads to detachment from reality—not unlike Barbara Bush's comment that she doesn't want to trouble her "beautiful mind"[56] with statistics about troop or civilian casualties in Iraq. It's all so American/Judeo-Christian—and, of course, New Age: keeping a positive attitude so that we never feel badly about what's actually happening.

How unfortunate, according to such an attitude, that someone like me would ask readers to feel the depths of their grief, fear, anger, or despair about the death of the planet and its inhabitants and talk and work with other humans to prepare for collapse. A righteous attitude bypasses those emotions and makes the state of our planet someone else's problem, not *my* problem. It communicates that one is above emotions and really doesn't want to soil his/her sanitized psyche with them.

The addiction to a "positive attitude" in the face of the end of the world as we have known it is beyond irrational. It's an obsession that could only be cherished by humans; it is, indeed human-centric, as if human beings are the only species that matter and as if the most crucial issue is that those humans are able to feel good about themselves as the world burns.

Usually, having a "positive" attitude about collapse implies wanting it not to happen, believing that it may not happen, and doing everything in one's power to convince oneself that it won't happen. This is a uniquely human attitude. If we could interview a polar bear who had just drowned trying to find food because the ice shelves that he usually rested on which allowed him to regain his strength during the hunt were no longer there, I suspect he'd express a very different attitude.

Now of course, we have the delusional human element who argue that humans are not killing the planet—as if the hairy-eared dwarf lemur, the pygmy elephant, or the ruby topaz hummingbird were responsible. Who else has skyrocketed ocean acidity to exponential levels, who else is inundating the atmosphere with carcinogens, turning topsoil into sand which contains as many nutrients as a kitchen sponge, and is rapidly eliminating clean, drinkable water from the face of the earth?

Derrick Jensen in *Endgame,* Volume I, states that "The needs of the natural world are more important than the needs of any economic system." (127) He continues:

Any economic system that does not benefit the natural communities on which it is based is unsustainable, immoral, and really stupid.(128)

Explaining human disconnection from the rest of earth's inhabitants, Jensen describes the various layers of resistance among humans to their innate animal essence. One of the deeper layers is our "fear and loathing of the body", our instinctual wildness and therefore, our vulnerability to death which causes us to distance ourselves from the reality that we indeed are animals.

Perhaps if you journaled about this at the end of Chapter 16, "What The Creatures Can Teach Us", you became aware of this very phenomenon. In fact, it is one of civilization's fundamental tasks. Have not all modern societies disowned and genocided the indigenous? And for what purpose? Not only for the purpose of stealing their land, eradicating their culture, and eliminating so-called barriers to "progress", but because native peoples (you know, "savages") as a result of their intimate connection with nature, are such glaring reminders of humankind's animal-ness. They are embarrassingly "un-civilized." Thus, modernity must "civilize" the savage in order to excise the animal, inculcating in him a human-centric world view.

The consequence has been not only the incessant destruction of earth and its plethora of life forms, but the murder of the human soul itself. Benjamin Franklin said it best after returning from living with the Iroquois: "No European who has tasted Savage life can afterwards bear to live in our societies."[57]

Any person who wants to "maintain a positive attitude" in this culture—the culture of civilization that is killing the planet—killing all the people and species that we all love—that person is not only irrational and deeply afflicted with denial, but he is exactly like a member of an abusive family system in which physical and sexual assault are occurring in the home on a daily basis, but that family member insists on "thinking good thoughts" and resents anyone and everyone who says what is so about the abusive system.

So let's admit two things: 1) Humans are fundamentally animals. Yes, we are more than animals, but civilization with its contempt for the feral has inculcated us to own the "more than" and disown everything else. 2) The culture of civilization is inherently abusive, and it is abusive precisely because it has disowned the animal within the human. Indeed animals kill other animals for survival, but they do not soil, conquer, rape, pillage, plunder, enslave, pollute, slash, burn, and poison their habitat—unlike those "more-than-animal" beings who seem incapable of *not* doing all of the above. Conversely, the "more-than-human" creatures respect their surroundings because they instinctively sense that their survival depends on doing so.

We insist that we are more intelligent than the more-than-human world, but a growing body of evidence undermines that assumption. A 2007 Japanese study[58] revealed that when young chimps were pitted against human adults in two short-term memory tests, overall, the chimps won. Researcher Tetsuro Matsuzawa of Kyoto University said that the study challenges the belief that "humans are superior to chimpanzees in all cognitive functions."

Moreover, a British study[59] at the University of St. Andrews confirmed that elephants keep track of up to 30 absent relatives by sniffing out their scent and building up a mental map of where they are. Herd members use their good memory and keen sense of smell to stay in touch as they travel in large groups, according to a study of wild elephants in Kenya. Dr. Richard Byrne of St Andrews noted that elephants have two advantages over humans - their excellent sense of smell and, if their popular reputation is anything to go by, a good memory.

One may argue that neither a chimp nor an elephant could design a computer, but I ask: What is more consequential, the ability to design a computer or the ability to protect, sustain, and nurture the planet on which one resides? Of what value is the computer if none of us is here to use it?

Civilization, which has never ceased soiling its nest since its inception, has also never understood its proper place on the earth: that of a guest, a neighbor, a fellow-member of the community of life. As a result, everything civilization has devised and which is "unsustainable, immoral, and stupid", as Jensen names it, is now in the process of collapsing.

I ask for an honest answer here: How can anyone tell me with a straight face (or a righteous attitude) that that reality is "negative"? Would the seagull on a Southern California beach with her feet entangled and bleeding in plastic netting left behind by "more-than-animal" life forms tell me that the collapse of what created her plight is "negative"? Would thousands of dead spruce trees in Colorado ravaged by beetles as a direct result of climate change tell me that collapse is a bad idea? Would the plankton and bleached coral at the bottom of the sea which are fading and dying with breathtaking rapidity as a result of global warming, tell me to keep a positive attitude and do everything in my power to stop the collapse of civilization? I think not.

Fundamentally, what all forms of positive thinking about collapse come down to is our own fear of death. Thanks to civilization's Judeo-Christian tradition and its other handmaiden, corporate capitalism, humans have become estranged from the reality that death is a part of life. Human hubris gone berserk as a result of a tumescent ego, uncontained by natural intimacy with the more-than-human world, believes humanity to be omnipotent and entitled to invincibility. Therefore, from the human-centric perspective "collapse should be stopped" or "maybe it won't happen" or "somehow humans will come to their senses". Meanwhile, the drowning polar bears inwardly wail for the death of humanity as the skeletons of formerly chlorophyll-resplendent Colorado spruce shiver and sob in the icy December wind. Alas, our moral, spiritual, and human obligation is to discard our positive attitude and start feeling their pain. Until we do, we remain human-centric and incapable of seizing the multitudinous opportunities that collapse offers for rebirth and transformation of this planet and its human and more-than-human inhabitants.

May I remind us: We are *all* going to die. Or as Derrick Jensen writes in *Endgame*:

> *The truth is that I'm going to die someday, whether or not I stock up on pills. That's life. And if I die in the population reduction that takes place as a corrective to our having overshot carrying capacity, well, that's life, too. Finally, if my death comes as part of something that serves the larger community, that helps stabilize and enrich the landbase of which I'm part, so much the better.*

Now, I hasten to add that I am not suggesting we select our most "negative" emotion about collapse, move in, redecorate, and take up residence there. Feel one's feelings? Yes, and at the same time revel in those aspects of one's life where one feels nourished, loved, supported, comforted, and in those people and activities that give one joy and meaning.

Had civilization not spent the last five thousand years attempting to murder the indigenous self inherent in all humans, we would not have to be told, as native peoples and the more-than- human world do not, that most of the time, life on this planet is challenging, painful, scary, sad, and sometimes enraging. What our indigenous ancestors had and still have to sustain them through the dark times was ritual and community. Our work is to embrace and refine both instead of intractably clinging to a "positive attitude" in the face of out-of-control, incalculable abuse and devastation.

In his article "The Planned Collapse Of America"[60], Peter Chamberlin asserts that a small group of ruling elite has been engineering the economic and social collapse of the United States for some time. While I agree and also fear the economic meltdown and social and political repression to which Chamberlin alludes, his perspective is once again, human-centric and Amero-centric. Reality check: Collapse is indeed happening, but it is occurring globally and threatening to annihilate all nations and all species. *That* collapse was not "planned" by ruling elites, and it is one in which all humans have participated. It now has a life of its own and is most likely, out of our control. Attempting to abort it or blame other humans for it is a waste of time and energy.

As I stated above, the question for humans is not: What do we do *about* collapse? but rather, What do we do *with* it? It holds inestimable opportunities for rebirth and intimacy with other humans and the earth community, but only if we open to it. Opening to it means opening to our own mortality, which as Derrick Jensen insists, may be part of something that serves the larger community. Perhaps one opportunity collapse is presenting us is that of moving beyond our human-centric perspective— our hubris and addiction to invincibility, begging us to humble ourselves and crawl behind the eyes of the more-than-humans as Joanna Macy poignantly writes:

> *We hear you, fellow-creatures. We know we are wrecking the world and we are afraid. What we have unleashed has such momentum now; we don't know how to turn it around. Don't leave us alone; we need your help. You need us too for your own survival. Are there powers there you can share with us?*

Indeed there are powers they can share with us, but not until we can let go of our current definition of "positive" and, feeling their pain, finally comprehend that the collapse of civilization may be the best thing that could happen to all of us.

While much is spoken and written here and elsewhere about "surviving collapse", the reality is that not everyone will. Furthermore, the hardiest survivors are nevertheless mortal. The ultimate "collapse", in the form of physical death, lies somewhere ahead of us all. In the kinds of communities that most individuals preparing for collapse are envisioning, death will be a part of life, but hopefully, with far more transparency, honesty, community support, and conscious intention than a death-denying civilization has allowed.

As noted in an earlier chapter, we more skillfully prepare ourselves for the "big" death when we can open to the "little" deaths in our lives. Also mentioned was the notion of functioning collectively as hospice workers for the world, but of course, that does not preclude the likelihood of being literal hospice workers for loved ones making their transition from this life. Our obligation to the dying is to be present with them, but in order to do so, we must be present with the reality of our own death. How do we do that? What does "being present" mean anyway?

As Mary Oliver's poem above paradoxically emphasizes, those who are most present with their own death are those who are the most fully, vibrantly alive. Anything less is somnambulism, or sleep walking—a malady with which the culture of empire is severely and widely afflicted.

Chapter One of this book opened with Jared Diamond's assertion that the human race is committing suicide, and not only killing itself but everything in its domain. The rest of this book has been somewhat of a manual for how not to commit suicide and how to consciously prepare on a variety of levels for the suicidal death of civilization, in other words, attending to the process of choosing life and connection over death and estrangement. Yet we are all mortal, and regardless of how valiantly and robustly we may survive collapse, we must come to terms with the final limitation, death.

At times I feel almost incapable of comprehending the magnitude of death that collapse will visit upon the inhabitants of planet earth, and as I have noted above, the former subtitle of this book "restoring life on a dying planet" was not chosen to soften the word "collapse" in the title. The human and non-human worlds are terminally ill, but remaining conscious before, during, and after collapse means that we do not have to commit suicide, but neither do we need to die mindlessly as victims of something that "just happened" to us. When death approaches, our work—our most daunting challenge is to consciously claim it because an integral part of restoring life is the willingness to be present with death. Stephen Levine notes in "Opening to Death" that "In the American Indian wisdom, wholeness is not seen as the duration one has lived but rather the fullness with which one enters each complete moment."[61]

The Dali Lama's practice of spending one hour each day pondering his own death is ludicrous—beyond absurd, in the culture of empire, invested as it is in heroically achieving and deluding itself that it is somehow immortal and invincible. Yet the Dali Lama's daily contemplation of death appears to have made him even more alive than he might otherwise be. No doubt his or anyone's contemplation of death is facilitated and enriched by noticing what also is attempting to be born as a result of the death at hand. The awareness of the latter does not minimize the grim finality of death, but rather, enhances its meaning in our lives. *Attending consciously to one's own death in the throes of vibrant aliveness is a practice that allows one to be supremely present with death.*

Being present simply means that we stay with, rather than flee—that we are available emotionally, physically, mentally, and spiritually. One place to begin the process might be to notice that right now, as you read these words—attentively, breathing, noticing your bodily sensations and emotional reactions, you are also dying. It may be useful to take a moment to notice what you feel as you consider that reality. Consciously integrating the reality of death with one's aliveness is arduous and challenging work, but echoing the wisdom of Mary

Oliver's poem, Levine reminds us: "There seems to be much less suffering for those who live life in the wholeness that includes death. Not a morbid preoccupation with death but rather a staying in the loving present, a life that focuses on each precious moment. I see few whose participation in life has prepared them for death. Few who have explored their heart and mind as perfect preparation for whatever might come next be it death or sickness, grief or joy."

Echoing Levine's perspective regarding death is that of Malidoma Somé who addresses death in relation to the community and emphasizes that connection with the community of life, as well as one's immediate community, cannot be broken:

> For the Dagara people, death results in simply a different form of belonging to the community. It is a lesson from nature that change is the norm, that the world is defined by eternal cycles of decline and regeneration. Having journeyed adequately in this world in your life, you become much more effective to the community that contained you when you return to the world of Spirit. When my grandfather, Bakhyè, died, he told my father, "I have to go now. From where I'll be, I'll be more useful to you than if I stay here." Death is not a separation but a different form of communion, a higher form of connectedness with the community, providing an opportunity for even greater service.[62]

Macrocosmically speaking, the collapse of civilization is a death that will ultimately annihilate attitudes, lifestyles, and human patterns that have been killing off all of earth's life forms for centuries. Alberto Villoldo states in "2012: The Odyssey" that we must come to grips not only with our own mortality but with the extinction of the human race. Although we may shudder when we hear of the extinction of 200 species per day now occurring on this planet, can any of us fully comprehend the extinction of the human race? Villoldo is not forecasting the certainty of human extinction, but only imploring us to consider the possibility.

Although it may require several million years, the earth is capable of regenerating and bringing forth abundant life, but apparently that will not be possible apart from the destruction of civilization. Microcosmically, our work as individuals and communities is to notice what might be born in current time as a result of our own personal deaths. What might our transition bring to life, not only for ourselves but for our loved ones? And as we assist others in dying, we must ask similar questions. What rituals are being called for? What does the life—and the death—of the departing or departed one have to teach us?

All of the issues I have raised in this chapter are ones that the culture of civilization overall, and most of the individuals within it, cannot bear to entertain. This is precisely why anyone who speaks consciously and honestly about collapse is generally met with either intractable resistance or somnambulistic apathy.

So I ask: Must a culture that will not grapple with a fundamental aspect of life, namely death, be forced to do so by the consequences of suicidal scenarios it has been busy orchestrating for the past 5,000 years? As I dialog with my fellow earthlings who are aware of and preparing for collapse, I notice that without exception, they are individuals who are capable of contemplating their own and civilization's death. Their lifestyles perpetuate neither hubris nor heroics. They can talk about collapse and attend to the business of survival precisely *because* they realize they may perish! Most of them have a connection to something greater and more mysterious than themselves; most have made friends with "defeat" as echoed in Gibran's poem in Chapter Four. Ironically, I do not find many among us who are morbid souls, clinically depressed, or what the culture of empire would call "losers". We are all wounded, but the majority of us are also passionately alive, doing the work in the world that we feel called to do.

Collapse holds and will hold many different things for everyone, yet for those who are awake to it, it feels like a protracted process of letting go, opening, surrendering to a new lifestyle, new friends, a new community, a new residence, new work, and of course, new and unprecedented challenges. As we embrace the new we also learn to release the old and allow what is not necessary or useful to fall away. At times this is frustrating, stressful, frightening, sad, painful--even excruciating, and at times it feels like a long-overdue liberation that we cannot wait to celebrate. Thus, collapse is both an external process and one that is occurring internally through the dissolution of ego defenses and patterns of survival that may have served us well in the culture of empire but may spell our doom as civilization crumbles around us.

Our work is two-fold—first, to embrace the "little deaths" of life so that we may cultivate the needed grace to confront the "big death." Elizabeth Bishop's famous poem, "One Art" is instructive and inspirational:

> *The art of losing isn't hard to master;*
> *so many things seem filled with the intent*
> *to be lost that their loss is no disaster.*
>
> *Lose something every day. Accept the fluster*
> *of lost door keys, the hour badly spent.*
> *The art of losing isn't hard to master.*
>
> *Then practice losing farther, losing faster:*
> *places, and names, and where it was you meant*
> *to travel. None of these will bring disaster.*
>
> *I lost my mother's watch. And look! my last, or*
> *next-to-last, of three loved houses went.*
> *The art of losing isn't hard to master.*

> *I lost two cities, lovely ones. And, vaster,*
> *some realms I owned, two rivers, a continent.*
> *I miss them, but it wasn't a disaster.*
>
> *--Even losing you (the joking voice, a gesture*
> *I love) I shan't have lied. It's evident*
> *the art of losing's not too hard to master*
> *though it may look like (Write it!) like disaster.*

And as we work with opening to the "little deaths" we may find extraordinary courage and resources within which will facilitate our opening to the end of our physical existence. Collapse is here to bring us "defeat" and "death", but also the possibility of more passion, aliveness, courage, compassion, and wholeness than we may have ever found possible in the pernicious, soul-murdering milieu of civilization.

RELFECTION

Journaling or drawing are highly recommended for working with these questions

** What makes you feel most alive? What is your passion?

**Is your body fully alive? If not, what needs to happen for you to feel physically vibrant?

**Do you have access to your emotions? Do you allow yourself to feel grief? Anger? Fear? Joy? What barriers exist within you that may impede your feeling any of these feelings? Take some time for journaling or drawing about this.

**How does your spiritual path and/or your connection with nature assist you in feeling alive and energetic?

**How much have you thought about your own death? What are your fears regarding death? What aspects of death feel less fearful?

**Illness and disability are reminders of our mortality. Do you have a serious illness or physical limitations? If so, it may be useful to journal or draw images connecting the illness or limitation to your own mortality.

**Have you ever had a terminal illness or a near-death experience? If so, what have you learned from it?

**What does the phrase "opening to death" mean to you? What kinds of "little deaths" might you need to open to?

**Who would you like to have with you when you die? How would you like them to be present to you?

**What is your experience of being present to others in death? What have you learned from the experience(s)?

**In awakening to collapse what within you has died? What have you released in your inner world? What have you released in your outer world? What do you still resist releasing?

**Pleaseseenextpageforaspecificjournalingexercise entitled "How I Wish To Be Remembered"

EXERCISE: HOW I WISH TO BE
REMEMBERED

The following exercise appears in *The Four Insights: Wisdom, Power, and Grace of The Earthkeepers*, by Alberto Villoldo[63]. It is a powerful and nurturing process which enables one to consciously attend to one's own death and at the same time, deepen the vibrance of one's current life.

Imagine that you've lived a long and rich life, and now you're on your deathbed. Write your own eulogy, featuring rich detail about how you lived, how you loved, what adventures you went on, how you were of service, and how you wish to be remembered. How did you touch other people? What did you learn? What did you overcome? What meant the most to you?

When you've finished writing this eulogy, you might want to share it with those you love, as it is a road map for the life you are called to live but may not be living. Think about whether you're actually on the road to living this life that you've described; if not, ask yourself what needs to change today.

NOTES

NOTES

Chapter 20—Bringing Back the World

A Valley Like This

Sometimes you look at an empty valley like this,
and suddenly the air is filled with snow.
That is the way the whole world happened -
there was nothing, and then...

But maybe sometimes you will look out and even
the mountains are gone, the world becomes nothing
again. What can a person do to help
bring back the world?

We have to watch and then look at each other.
Together we hold it close and carefully
save it, like a bubble that can disappear
if we don't watch out.

Please think about this as you go on. Breathe on the world.
Hold out your hands to it. When mornings and evenings
roll along watch how they open and close, how they
invite you to the long party your life is.

~William Stafford~

If you have arrived at this page after reading all of those above, I proudly salute you for your courage and perseverance. In so doing you have demonstrated that you desire to know the unvarnished truth regarding the state of the planet and its inhabitants and therefore have opened yourself to a plethora of emotions and insights. My deepest desire is that you are able to affirm that you have been significantly influenced by reading this book, but more importantly, that you will never cease living with the question of who you want to be in the face of collapse. If you are committed to pursuing the question, you will experience many occasions, undoubtedly many more than you would like, to continue working with it.

As I have stated here so many times, I do not know how collapse will unfold. I do not know who of us will survive it or if any of us will. I do not cherish fond hopes that collapse-watchers will create widespread networks of eco-villages with loving communities where vigilant sustainability, ingenious community currencies, or stupendous relocalization of economies prevail. I would consider the creation and maintenance of them after a global holocaust a breathtaking accomplishment. What is more likely is a civilization left in ruins with pockets of individuals or communities who have endured collapse, not only because they have been well-prepared, but because they have understood in the marrow of their bones the nature of the initiation to which they willingly opened. What they will be able to physically build or restore is anyone's guess. What *is* certain, however, is that for better or for worse, anyone who does survive the collapse of civilization will be a dramatically altered human being. Moreover, it may be that as a result of their transformation, they will be empowered and wizened on a cellular, soul level, to construct a radically new paradigm for humans, and that as a result of the experience through which they have lived, they will be the only living beings capable of "bringing back the world".

A marvelous manual for bringing back the world might be William Kotke's *The Final Empire: The Collapse of Civilization, The Seed of The Future.*[64] As part of my commitment to holding the tension of vision alongside current reality, I continue to cherish a particular line from Kotke's book in which he speaks of "gathering seeds of Natural cultures and the truly beneficial things created by civilization" and carrying them through the apocalypse. I can't think of a more succinct description of restoring life on a dying planet. More specifically, Kotke admonishes us to create seed communities amid and alongside accelerating collapse:

> *To be actively mobilizing toward setting up what might be called 'seed' communities is the really significant action. If people don't actually get out of the money economy to a significant degree, if they don't create a new land based culture that aids the earth, all the other political and environmental efforts will ultimately be meaningless.*

> *To be actively mobilizing toward setting up seed communities is what is most significant. Movement is now happening, the seed is being*

> *empowered. That we are moving toward food growing capability, land, community, emotional positivity, healing, integration of every level possible—and toward the top of the watersheds, that is the significant action—by whatever means we have at our disposal. Of course, people must resist the destruction and move ahead on all fronts that they are normally active in, but this becomes meaningless unless cultures of balance are also established.*

One specific and increasingly popular positive vision and workable model for "bringing back the world" is that of the Transition Town movement which originated in 2005 in the United Kingdom and which is now spreading around the world, literally like wildfire. I highly recommend reading Rob Hopkins' *The Transition Handbook: From Oil Dependency to Local Resilience*[65] which provides an extraordinary manual for individuals for organizing their communities in preparation for energy descent, food security, and climate change.

It has been with great pleasure and relief that I have looked deeply into the Transition Town movement and found it to exemplify everything that I believe comprises effective relocalization and the shaping of alternative economies and vibrant communities. I use the word "relief" because I am delighted that this movement is able to hold reality alongside one of its fundamental principles, resilience. To its credit, this book does not sugar-coat the daunting reality of Peak Oil and Climate Change, but rather, offers a positive vision of preparation and myriad practical steps for manifesting it.

One paragraph near the book's end suggests an integration of the Transition Movement with what you have read in this book:

> *While Peak Oil and Climate Change are understandably profoundly challenging, also inherent within them is the potential for an economic, cultural, and social renaissance the likes of which we have never seen. We will see a flourishing of local businesses, local skills and solutions, and a flowering of ingenuity and creativity. It is a Transition in which we will inevitably grow, and in which our evolution is a precondition for progress. Emerging at the other end, we will not be the same as we were: we will have become more humble, more connected to the natural world, fitter, leaner, more skilled, and ultimately, wiser.*

I believe that *Sacred Demise* and *The Transition Handbook* are necessary companions for navigating the waters of collapse, allowing us to hold in our hearts and minds the two complementary poles of, on the one hand, unabashed reality and, on the other, commitment of oneself to a positive vision. With all my heart, I want to support Transition Towns in my community and around the world with the hope that their implementations are not too little, too late. Yet, even if they are, I cannot think of a better place to direct one's energy, time, and passion--regardless of outcome, as we navigate with realism and resilience, the collapse of civilization. An overview of the Transition movement may be read at the organization's website.[66]

At the end of every semester in my classes, we spend at least four class periods examining collapse in depth. The final class period consists of a talking circle where we come together and discuss our feelings about collapse. Students unburden their souls about the impact the class has made on them—their fears, how the class has changed their minds about what career to pursue, how the class has altered their hopes and dreams about marriage and having children. We cry, we talk about our fears, we express anger, and we celebrate that we are all human beings who understand the reality of collapse and are sitting together in a circle talking about it.

As a college professor who is preparing consciously for the collapse of empire, I am well aware that public education on all levels is nearing extinction. Therefore, I see my role as an educator in terms of planting seeds regarding collapse awareness, with no ability to determine how those seeds will germinate or even if they will. Thus, I tell my students, sometimes tearfully, that I recall being their age as if it were yesterday. All I wanted to do was get good grades, have fun, and find the next party where I could blow off steam. But above and beyond all that, I wanted a future, and I wanted it to be successful, make lots of money, have a family, and feel like I was making a difference in the world. The last thing I would have wanted at their age was some silver-haired lady standing at the front of my history class telling me that I had no future—or that my future would look very different from the one I fantasized.

I tell my students that I know they will leave my class, take many other classes where they will hear nothing about collapse but only about how "rosy" their future is, and they will graduate, move out into the world and commence their adult professional lives, and most certainly, they will forget the history classes they had with me. How not? They are in their twenties and bursting with hopes and dreams. How could I expect them to keep something like collapse on the front burner of their awareness?

But I also tell them that somewhere, somehow— perhaps as they are standing in a check-out line paying $8 for a loaf of bread and $10 for a gallon of milk—if those items are even available--they will remember my class. Life will happen brutally and without mercy, and eventually, they will remember the silver-haired professor who told them how and why collapse might ruin the future of their dreams. My hope is that they will remember their sixteen weeks with me and that in some way, it will assist them in navigating the end of their world as they have known it. Perhaps their short time with me, all that they learned and felt and eventually needed to forget, will make it possible for them, in spite of themselves, to bring back the world.

But wait. It's 2009, and college students everywhere are beginning to awaken to the colossal delusion they have been sold in a society that tells them that if they graduate from college, they will secure a fabulous job, eventually achieve a six or seven-figure income, and live happily ever after. As one of my students recently opined, "I'm majoring in 'unemployment'. I am now working two jobs to pay for tuition so that I can graduate and then work two or three jobs to pay off my student loans. I'll end up a depressed workaholic like my father. What the hell am I doing in college?"

Like countless students who have entered the Second Great Depression, this young man has awakened and begun to question the American Dream which is a fantasy disguising an economic system of debt slavery. He sees through the illusion, but feels caught between a rock and a hard place because while he knows that a college education is not necessarily his best option, he hasn't yet discovered what is calling him, what his life purpose is, or how best to fulfill it. Furthermore, I suspect that within the next decade we are going to witness millions of students who abandon or are priced out of a college education who will be forced to confront, in the face of an economic depression that may surpass the severity of the depression of the 1930s, not merely what to do with their lives, but how to most creatively "bring back the world."

In my March, 2008 online review of Jim Kunstler's *World Made By Hand* I concluded my comments with essential questions that I believe we must consider when thinking about life in a post-collapse world— questions such as: "What will give us meaning? What will console us? What will allow us to keep going when any sense of purpose has eluded us?" In response to these questions, one reader of my article, who had read the book and contemplated the author's description of life in such a world, stated:

> *I, for one, would find much more meaning from putting food on the table that is truly needed and sustaining rather than taken for granted. Food that I raised or killed myself, or we ourselves, or my neighbor did, and I bartered with him for it. Much more so than the meaning Empire tells me what I am supposed to get from sitting here in my cubicle (my penultimate day today!) rearranging little electronic blips in exchange for money, which I am then supposed to exchange not only for my sustenance, but also for all sorts of diversions, to make me forget how meaningless it all is.*

I, for one, will find consolation in knowing my neighbors – and in knowing that they are there for me as I am for them, rather than living amidst strangers, as most all of us do now. I will find consolation in knowing that my ecological footprint does not extend beyond my gaze. That the things I consume do not cause death and destruction beyond my ability to see and internalize, rather than out of sight and mind as now, and so much larger than any being could ever have a 'right' to.

I, for one, will find purpose in working closely and cooperatively and communally with those around me to provide our own sustenance, comforts such as they may be, and entertainments as time allows.

*I have no illusions that life post-collapse will be idyllic, nor that the transition will be anything but ugly. But neither shall I miss that which is dying – the dizzying complexity of our oil-drenched lifestyles, a thousand channels of nothing worth watching, mega-malls, motor sports (how many kinds of insane are those!?!), celebrities, glitter, growth, more, faster, bigger, keep up with the Joneses but ignore the sweatshops and the dying ecosystems, consume, medicate, live large... then die. Where is one to find a sense of purpose in all of **that**?*

This reader's response, in just a few short paragraphs poignantly and succinctly captures what I believe is the essence of the world that is now ending, as well as the one we may have the opportunity to re-create. No matter how loudly we may protest the culture of empire, it will not be as easy to walk away from as we may imagine. It may not have given us a just or humane existence, but it's predictable and familiar. Nor will allowing ourselves to be remade by a post-collapse world be anything but daunting. Nevertheless, if we are willing, like our indigenous ancestors, to engage in the struggle to liberate our bodies and souls from external and internal colonization, we may discover more fully our own deeper story and what it is that we came here to do.

The Thread

There is a thread you follow. It goes among
things that change. But it doesn't change.
People wonder about what things you are pursuing.
You have to explain about the thread.
But it is hard for others to see.
While you hold it you can't get lost.
Tragedies happen; people get hurt
or die; and you suffer and get old.
Nothing you do can stop time's unfolding.
But you don't ever let go of the thread.

~William Stafford~

SUGGESTED RESOURCES

<u>BOOKS</u>

David Abram

The Spell of the Sensuous: Perception and Language in a More-Than-Human World

Douglas Adams and Mark Carwardine

Last Chance to See

Sharon Astyk

Depletion and Abundance: Life on the New Home Front

A Nation of Farmers: Defeating the Food Crisis on American Soil

Carolyn Baker

Coming Out of Fundamentalist Christianity: An Autobiography Affirming Sensuality, Social Justice, and The Sacred

Reclaiming the Dark Feminine: The Price of Desire

US History Uncensored: What Your High School Textbook Didn't Tell You

Thomas Berry

The Dream of the Earth

The Great Work: Our Way into the Future

William R. Catton, Jr.

Overshoot: The Ecological Basis of Revolutionary Change

Diana Leafe Christian

Finding Community: How to Join an Ecovillage or Intentional Community

Julian Darley

High Noon for Natural Gas: The New Energy Crisis

Kenneth Deffeyes

Hubbert's Peak: The Impending World Oil Shortage

Beyond Oil: The View From Hubbert's Peak

Richard Douthwaite

The Growth Illusion: How Economic Growth has Enriched the Few, Impoverished the Many and Endangered the Planet

Short Circuit

David Edwards

Burning All Illusions

Charles Eisenstein

The Ascent of Humanity

Duane Elgin

Promise Ahead: A Vision of Hope and Action for Humanity's Future

Lyle Estill

Biodiesel Power

Small Is Possible: Life In A Local Economy

Robert Fritz

The Path of Least Resistance: Learning to Become the Creative Force in Your Own Life

Laurie Garrett

The Coming Plague: Newly Emerging Diseases in a World Out of Balance

John Taylor Gatto

Dumbing Us Down: The Hidden Curriculum of Compulsory Schooling

Weapons of Mass Instruction

Chellis Glendinning

Off the Map: An Expedition Deep into Empire and the Global Economy

When Technology Wounds: The Human Consequences of Progress

My Name Is Chellis and I'm in Recovery from Western Civilization

Stan Goff

Full Spectrum Disorder: The Military in the New American Century

John Michael Greer

The Long Descent: A User's Guide to the End of the Industrial Age

David Ray Griffin

The New Pearl Harbor: Disturbing Questions About the Bush Administration and 9/11

Joan Halifax

The Fruitful Darkness: Reconnecting With the Body of the Earth

Thom Hartmann

The Last Hours of Ancient Sunlight: Waking Up to Personal and Global Transformation

Richard Heinberg

The Oil Depletion Protocol: A Plan to Avert Oil Wars, Terrorism and Economic Collapse

The Party's Over: Oil, War and the Fate of Industrial Societies

PowerDown: Options and Actions for a Post-Carbon World

Peak Everything: Waking Up To The Century of Declines

James Hillman and Michael Ventura

We've Had a Hundred Years of Psychotherapy and the World's Getting Worse

Rob Hopkins

The Transition Handbook: From Oil Dependency to Local Resilience

Derrick Jensen

A Language Older Than Words

The Culture of Make Believe

Strangely Like War: The Global Assault on Forests

Walking on Water: Reading, Writing, and Revolution

Endgame: Volume 1: The Problem of Civilization

Endgame: Volume 2: Resistance

Robert D. Kaplan

The Coming Anarchy: Shattering the Dreams of the Post Cold War

Penny Kelly

Robes: A Book of Coming Changes

David Korten

The Great Turning

William Kotke

Final Empire: The Collapse of Civilization and the Seed of the Future

James Howard Kunstler

The Long Emergency: Surviving the End of the Oil Age, Climate Change, and Other Converging Catastrophes of the Twenty-First Century

World Made By Hand

Paul Levy

The Madness of George W. Bush: A Reflection of Our Collective Psychosis

Grace Llewellyn

The Teenage Liberation Handbook: How to Quit School and Get a Real Life and Education

James Lovelock

The Revenge of Gaia

Mark Lynas

High Tide: The Truth About Our Climate Crisis

Six Degrees: Our Future on a Hotter Planet

Joanna Macy

World As Lover, World As Self

Despair and Personal Power in the Nuclear Age

Jerry Mander

In the Absence of the Sacred: The Failure of Technology and the Survival of the Indian Nations

Four Arguments for the Elimination of Television

Richard Manning

Against the Grain: How Agriculture has Hijacked Civilization

Michael Meade

The World Behind the World: Living At the End of Time

Jim Merkel

Radical Simplicity: Small Footprints On a Finite Earth

Deena Metzger

From Grief into Vision: A Council

Marc Crispin Miller

Loser Take All: Election Fraud and the Subversion of Democracy 2000-2008

Dmitry Orlov

Re-Inventing Collapse: The Soviet Example and American Prospects

Greg Palast

The Best Democracy Money Can Buy

Armed Madhouse

Stuart Pimm

The World According to Pimm: A Scientist Audits the Earth

Michael Pollan

Omnivore's Dilemma: A Natural History of Four Meals

In Defense of Food: An Eater's Manifesto

Daniel Quinn

Ishmael

The Story of B

My Ishmael

Beyond Civilization: Humanity's Next Great Adventure

Paul Roberts

The End of Oil: One the Edge of a Perilous New World

Marshall Rosenberg

Non-Violent Communication: A Language of Life

Barbara Rossing

The Rapture Exposed: The Message of Hope in the Book of Revelation

Michael Ruppert

Crossing the Rubicon: The Decline of the American Empire at the End of the Age of Oil

Presidential Energy Policy

Kirkpatrick Sale

Human Scale

Matt Savinar

The Oil Age is Over: What to Expect as the World Runs Out of Cheap Oil, 2005-2050

Jonathan Schell

The Fate of the Earth

Andrew Bard Schmookler

The Parable of the Tribes: The Problem of Power in Social Evolution

Malidoma Somé

Of Water and the Spirit: Ritual, Magic, and Initiation in the Life of an African Shaman

Ritual: Power, Healing, and Community

Eckhart Tolle

The Power of Now

A New Earth

Marshall Vian Summers

Greater Community Spirituality: A New Revelation

Steps to Knowledge: The Book of Inner Knowing : Spiritual Preparation for an Emerging World

Neale Donald Walsch

Conversations With God (and all books in the Conversations With God series)

Alexis Zeigler

Civil Liberty, Peak Oil, and the End of Empire

DOCUMENTARIES

9/11 Press For Truth

A Crude Awakening

The End of Suburbia

Escape From Suburbia

Hacking Democracy:

In Debt We Trust

Maxed Out

Money as Debt

The Power of Community: How Cuba Survived
Peak Oil

The Story of Stuff

The Truth and Lies of 9/11

Uncounted: The New Math of American Elections

Washington: You're Fired

What a Way To Go: Life At the End of Empire

Zero: An Investigation of 9/11

ARTICLES

Globalcorp, By Mike Ruppert

http://www.fromthewilderness.com/free/
ww3/031005_globalcorp.shtml

The "F" Word, By Mike Ruppert

http://www.fromthewilderness.com/free/
ww3/112001_f_word.html

Eating Fossil Fuels, By Dale Allen Pfeiffer

http://www.fromthewilderness.com/free/
ww3/100303_eating_oil.html

Coming Clean, By Catherine Austin Fitts

http://solari.com/archive/coming_clean/

The American Tapeworm, By Catherine Austin Fitts

http://solari.com/archive/the-american-tapeworm/

Narco Dollars For Beginners, By Catherine Austin Fitts

http://solari.com/articles/scoop_narco_dummies.htm

Beyond Hope, By Derrick Jensen

http://www.orionmagazine.org/index.php/articles/article/170/

Catastrophe As Spiritual Practice, By Sally Erickson

http://www.whatawaytogomovie.com/2007/05/13/catastrophe-as-spiritual-practice/

Eco-Anxiety Retrospective

http://eco-anxiety.blogspot.com/

ENDNOTES

[1] John Michael Greer, *The Long Descent*, New Society, 2008.

[2] http://www.context.org/ICLIB/IC34/Roszak.htm

[3] Thomas Moore, *The Re-Enchantment Of Everyday Life*, Harper Collins, 1996, p. 44.

[4] *Ibid.*, p. 303.

[5] *Ibid.*, p. 340

[6] Malidoma Somé, *The Healing Wisdom Of Africa*, Tarcher/Putnam, 1998, p. 28.

[7] Bill Plotkin, "Groundwork", http://www.selfgrowth. com/articles/Plotkin1.html

[8] http://trumpeter.athabascau.ca/index.php/trumpet/ article/view/407/658

[9] Charles Eisenstein, *Ascent Of Humanity*, Panthea Press, 2007, p. 236.

[10] http://www.whatawaytogomovie.com/2007/11/13/ build-an-ark-build-it-now/

[11] http://network.bestfriends.org/newhampshire/ news/19173.html

[12] David Korten, *When Corporations Rule The World*: http://www.amazon.com/When-Corporations-World-David-Korten/dp/1887208046

[13] http://carolynbaker.net/site/content/view/883/1/

[14] Richard Heinberg, *Peak Everything*, http://www.amazon.com/Peak-Everything-Century-Declines-Publishers/dp/086571598X/ref=pd_bbs_sr_1?ie=UTF8&s=books&qid=1196713162&sr=1-1

[15] http://www.hackingdemocracy.com/

[16] http://www.uncountedthemovie.com/about-the-film.html

[17] Marc Crispin Miller, *Loser Take All: Election Fraud and the Subversion of Democracy, 2000-2008*, IG Publishing, 2008.

[18] Thomas Berry, *The Dream Of The Earth*, p. xii.

[19] http://www.truthout.org/docs_2006/091407K.shtml

[20] http://www.context.org/ICLIB/IC34/Some.htm

[21] http://www.sourcewatch.org/index.php?title=Blackwater_USA

[22] http://www.usatoday.com/news/nation/2007-09-13-backpack_N.htm

[23] Bill Plotkin, *Nature And The Human Soul*, New World Library

[24] Report on America, International Psychoanalytic Congress, Nuremberg, 1910

[25] *Comfortable With Uncertainty*, Pema Chodron, p. 47.

[26] http://carolynbaker.net/site/content/view/53/3/

[27] Thomas Moore, *The Re-Enchantment Of Everyday Life*, p. 4.

[28] *Ibid.*, p. 5.

[29] http://ecopsychology.athabascau.ca/Final/duncan.htm

[30] http://www.context.org/ICLIB/IC34/Roszak.htm

[31] David Abrams, *The Spell Of The Sensuous*, Vintage, 1997, p. 268.

[32] Alberto Villoldo, *The Four Insights: Wisdom, Power and Grace of the Earthkeepers*, "The Query", P. 190

[33] http://www.randypeyser.com/levine.htm

[34] John O'Donohue, *Beauty: The Invisible Emrace*, http://www.amazon.com/Beauty-Invisible-Embrace-John-ODonohue/dp/B000GH2YUQ/re

f=sr_1_1?ie=UTF8&s=books&qid=1196715558&s
r=1-1

[35] *Dmitry Orlov, Re-Inventing Collapse: The Soviet Example And American Prospects, 78-79.*

[36] http://www.livingeconomies.org/

[37] http://www.fromthewilderness.com/free/ww3/060105_soviet_lessons.shtml

[38] http://www.cnvc.org

[39] http://www.solonline.org/aboutsol/who/Senge/

[40] http://www.heartcircle.com/mf.html

[41] http://www.ojaifoundation.org/Content/council_training.php

[42] http://www.mscottpeck.com/

[43] http://women.timesonline.co.uk/tol/life_and_style/women/the_way_we_live/article1830703.ece

[44] http://carolynbaker.net/site/content/view/126/3/

[45] Native American Prophecy, http://www.2012endofdays.org/more/Native-American-prophecy.php

[46] Michael Meade, *The World Behind The World*, Greenfire Press, 2008

[47] http://news.independent.co.uk/sci_tech/article3075681.ece

[48] http://www.bestfriends.org/aboutus/staffdepartments/biommountain.cfm

[49] http://carolynbaker.net/site/content/view/185/

[50] http://peakoilmedicine.com/2007/03/

[51] http://www.energybulletin.net/21217.html

[52] http://www.permaculture.org/nm/index.php/site/index/

[53] http://www.hesperian.org/

[54] http://eco-anxiety.blogspot.com/2008/12/eco-anxiety-retrospective.html

[55] Angeles Arrien, *The Fourfold Way*, Harper, San Francisco, 1993, p. 113

[56] http://www.commondreams.org/views04/0429-11.htm

[57] http://www.ratical.org/many_worlds/6Nations/FFafterw.html

58 http://my.earthlink.net/article/top?guid=20071203/
47538d50_3ca6_1552620071203174788775

59 http://news.bbc.co.uk/2/hi/science/nature/7127276.
stm

60   http://www.bestcyrano.org/THOMASPAINE/
?p=472

61 http://www.thebody.com/content/living/art14052.
html

62 Malidoma Patrice Somé, *The Healing Wisdom of Africa*, p. 53.

63 Alberto Villoldo, *The Four Insights*, p. 106.

64 William Kotke, *The Final Empire: The Collapse of Civilization, The Seed of The Future*, Author House, 2007.

65 Rob Hopkins, *The Transition Handbook: From Oil Dependency to Local Resilience*, Chelsea Green, 2008

66   http://transitiontowns.org/TransitionNetwork/
TransitionPrimer

ABOUT THE AUTHOR

Carolyn Baker, Ph.D., is an adjunct professor of history and psychology and manages her website, Speaking Truth to Power at www.carolynbaker.net. She is the author of *U.S. History Uncensored, Coming Out of Fundamentalist Christianity,* and *The Journey of Forgiveness,* as well as numerous articles on issues of environmental and social justice, consciousness, emotional and spiritual well being. She lives and works in Vermont.

LaVergne, TN USA
15 April 2010
179358LV00007B/14/P

9 781440 119729